T0342791

# Quarterly Essay

## CONTENTS

Quarterly Essay is published four times a year by Black Inc., an imprint of Schwartz Media Pty Ltd. Publisher: Morry Schwartz.

ISBN 978-1-86395-552-2 ISSN 1832-0953

Subscriptions – 1 year (4 issues): $49 within Australia incl. GST. Outside Australia $79. 2 years (8 issues): $95 within Australia incl. GST. Outside Australia $155.

Payment may be made by MasterCard or Visa, or by cheque made out to Schwartz Media. Payment includes postage and handling.

To subscribe, fill out and post the subscription card or form inside this issue, or subscribe online:

**www.quarterlyessay.com**
subscribe@blackincbooks.com
Phone: 61 3 9486 0288

Correspondence should be addressed to:

The Editor, Quarterly Essay
37–39 Langridge Street
Collingwood VIC 3066 Australia
Phone: 61 3 9486 0288 / Fax: 61 3 9486 0244
Email: quarterlyessay@blackincbooks.com

Editor: Chris Feik. Management: Sophy Williams, Caitlin Yates. Publicity: Elisabeth Young. Design: Guy Mirabella. Assistant Editor/Production Coordinator: Nikola Lusk. Typesetting: Duncan Blachford.

Printed by Griffin Press, Australia. The paper used to produce this book comes from wood grown in sustainable forests.

# MAN-MADE WORLD

## Choosing between Progress and Planet

### Andrew Charlton

## WHAT WE LEARNED IN COPENHAGEN

At 10.45 p.m. my phone rang. "The Danes are switching to the back-up plan," a voice said. "Room 20. 11.30 tonight." I pulled on my suit jacket, took up my warm coat and ventured downstairs into the night. I followed the now familiar route through Copenhagen's cobbled laneways and grand boulevards, across the bascule bridge straddling its narrow harbour straits, arriving finally at the United Nations conference venue on Amager Island. The meeting place was a small, windowless room on the second floor. I took one of the seats behind a folded paper sign with "AUSTRALIA" printed on both sides.

The final hours of the Copenhagen Climate Conference unfolded in that room. By midnight two dozen presidents and prime ministers from the world's most powerful nations had assembled with their advisers. Of course, this was not what the leaders had expected when they flew into the Danish capital for the second-last day of the two-week conference. They were scheduled to attend a state dinner at Christiansborg Palace, deliver a prepared speech on the conference floor and shake hands on a climate deal already hammered out by their negotiators. But as their jets touched down on the icy runway, weary officials greeted them with the news: there was no deal. Lengthy negotiating sessions had dissolved into protests, bickering and brinkmanship. Global leaders had flown into a failing summit and a looming political disaster.

The Danish hosts had convened the midnight meeting as a desperate last-ditch measure. Convinced that the formal negotiating process was hopeless, the Danes gambled that a small group of key leaders might be able to break the stalemate. Cramped and airless, the room lacked the opulence and technical wizardry typical of global summits. There was no lavishly adorned conference table for the leaders. No concentric circles of advisers, attendants and security. No simultaneous translators mouthing like goldfish inside soundproof glass booths. Instead, leaders hunched in plastic chairs around a rectangle of contiguous small tables. The scene looked more like a crowded primary-school classroom than a global summit freighted with the hopes of the world.

Barack Obama flew into Copenhagen later than the other leaders. When he reached the meeting at around 9 a.m., his presence injected a much-needed optimism into the room. The faces of frustrated leaders and their wilting advisers instantly brightened, as if salvation had just opened the door. Obama's personal staff sometimes jest that they work for "Black Jesus." Suddenly I got the joke. Everywhere he goes, people expect a miracle. But a miracle was not looking likely. As Obama took his seat, his Secretary of State, Hillary Clinton, summarised the state of play. "Mr President," Clinton said, "this is the worst meeting I've been to since the eighth-grade student council."

For the next twenty hours the room whirred with debate and dispute. Negotiations continued through the day, sometimes in the room, sometimes spilling out into the corridor, and at one point in the Chinese premier's hotel on the other side of town. Rich countries offered more financial support and greater flexibility for poor countries. Momentum surged forward and fell back. A deal seemed won and then lost. Finally, long after we had begun, the meeting dissolved in substantial disagreement. There had been no miracle. A short face-saving statement was all that was salvaged from the wreckage.

Thwarted and exhausted, leaders dispersed to face the bright glare and fading expectations of waiting news conferences. Ashen-faced under the lights of the television cameras, they delivered the bad news. Obama looked down the barrel of the camera and conceded that it was "not enough." Kevin Rudd – who with Penny Wong and her staff in the Department of Climate Change had worked as hard as anyone to get a deal – admitted he was "disappointed."

Elsewhere the disconsolate began pointing fingers. Green activists flung insults at departing delegates. "The city of Copenhagen is a crime scene tonight," Greenpeace seethed, "with guilty men and women fleeing to the airport." In the press centre, journalists sharpened their knives for a bloodletting across the world's front pages. A British editor inked the next day's headline: "A historic failure that will live in infamy." Sydney's *Daily Telegraph* prepared to splash with a bitter pun: "COP OUT."

When, after the failure of negotiations, we left the convention centre at 4 a.m., I'd been awake for forty-five hours. Many of the Australian team had been awake for longer, an exhausting sprint at the end of a marathon year of preparation. Through the haze of fatigue and crushed expectations there was one outstanding question: how could this have happened? Billed as the "summit to save the world," the Copenhagen Climate Conference had attracted unprecedented global attention. Decades of scientific work had built the case for action. More than a hundred heads of state, thousands of negotiators and countless NGOs had converged. How, with the best of intentions, with the world watching, could something so important go so wrong?

There was no shortage of answers, but few of them were satisfying. Many Western commentators immediately accused their political leaders of venality and sabotage; one British journalist reported that a deal was "systematically vetoed by the governments of North America and Europe." The environmental campaigner George Monbiot blamed American oil companies: "This [failure] is the result of a systematic campaign of sabotage by certain states, driven and promoted by the energy industries." Much of this didn't ring true to me because, as far as I could tell, Western leaders were pushing hardest to get a deal. At one point, an obviously frustrated Angela Merkel had demanded, "Why can't we even mention our own targets?" when poor countries rejected her attempt to record strong commitments that would have applied only to rich countries. Every Western leader in that room was seeking to strengthen the deal, not weaken it.

Others blamed China. Ed Miliband, then Britain's climate secretary, claimed that tougher emissions reductions "were vetoed by China, despite the support of a coalition of developed and the vast majority of developing countries." Again this doesn't feel like the full story. No single country was powerful enough to veto the deal without support from a wider constituency. Even the United Nations came in for criticism for its organisation of the official meetings. It is true that the UN's ham-fisted attempts at democracy had delivered anarchy on the conference floor, but the disorder and procedural bickering wasn't the root cause of the deadlock; the chaos was merely a symptom of a deeper discord.

The fundamental problem was not the United States, Europe, China or even the United Nations. Accusations of "cowardice," "stupidity," "mendacity" or any other simplistic tabloid insult missed the point. The deal broke down because Copenhagen exposed the central dilemma of our century: the choice between progress and planet.

Our planet is home to 7 billion people. Of these, roughly 1 billion live in rich countries: North America, Europe, Japan and Australia. The leaders of these countries arrived in Copenhagen persuaded of the urgency of the environmental challenges facing our planet. Backed by thousands of journalists and green activists, they pushed for a strong global agreement.

Another 6 billion people on our planet live in developing countries. Two billion of these – mainly in Africa and South Asia – are so poor that they barely have enough food to eat. Developing countries arrived in Copenhagen with their own priorities. Poor countries care about the environment, but poverty is their chief concern. A Chinese official made the point starkly: "You cannot tell people who are struggling to earn enough to eat that they need to reduce their emissions." Developing countries were unwilling to accept any binding constraints on their path out of poverty. "For centuries your countries have prospered by exploiting the world's resources," a Latin American negotiator explained to me. "How can I tell the slum dwellers they must stay poor to help clean up your mess?"

This was the conundrum in Copenhagen. A fraction of the world's people had become rich by plundering our planet to the point of exhaustion; now the still-poor majority wanted to do the same. "We shouldn't be too hard on ourselves," a German colleague said to me. "We have to be realistic about the problem. The world is split between those who want to save the planet and those who want to save themselves."

This essay is about that split: rich versus poor; planet versus progress. Copenhagen was just one global summit, but it was a symbolic battle in a broader conflict between economics and the environment. That conflict is defining the most important choices facing Australia and the world in the twenty-first century.

## Progress and planet

Humans have been on the earth for a tiny fraction of geological history, but in that short time we have come to dominate the planet. The natural world bears the mark of human activity more deeply and widely than that of any species before us. Today more than one-third of the earth's habitable surface is dominated by crops and human settlements, and a further third supports grazing animals. Expanding cities gobble up their hinterlands, forests are felled for farmland, and millions of miles of roads cut across continents. In all, nearly four-fifths of the planet's ecosystems are under human influence. "Give me a Wildness whose glance no civilization can endure," wrote Henry David Thoreau. If the Yankee poet was right that "in Wildness is the preservation of the World," then our source of salvation is ebbing away. Human civilisation has tamed nature, confining "wildness" to a shrinking corner of the globe.

Our species consumes a staggeringly disproportionate share of the world's resources. Measured by weight, we make up less than half of one per cent of the animals on the planet, yet we consume nearly 25 per cent of the total production of all land plants. Humans are using up the world's resources faster than any natural system can replace them. Scientists from the Global Footprint Network calculate that soon we will be extracting resources from the land and sea at twice the rate they are naturally replenished. At this rate, we are literally eating into our future.

In recent decades environmentalists have argued with mounting force that the growth of human activity on our planet is unsustainable. We are, they claim, on a collision course with destiny. "The world is in overshoot mode," says the environmentalist Lester Brown. "The environmental decline that will lead to economic decline and social collapse ... is well underway. No previous civilization has survived the ongoing destruction of its natural supports. Nor will ours." President Obama's science adviser, John Holdren, agrees: "we're driving in a car with bad brakes in a fog and heading for a cliff ... Prudence would suggest that we should start putting on the brakes." The Australian businessman and environmental commentator Dick Smith goes further: "we are not so much sleepwalking as sprinting towards the precipice."

But environmental warnings, dire as they may be, are not the only challenges we face. Indeed, these concerns can ring hollowly in the ears of poor countries, where environmental challenges are a distant threat compared to the daily tragedies of life in slums and villages. Across the world, 850 million people are malnourished. Nearly a billion people live in countries where the average income is less than $2 per day. Africa, in particular, is the epicentre of continuing human crisis. Desperate poverty, raging health pandemics and dirty water supplies have slashed life expectancy. Children are twenty times more likely to die in infancy in southern Africa than in Australia; those that survive can expect a life no longer, nor healthier, than that of an English person during the reign of Queen Victoria.

Poverty extends well beyond the least developed countries. Most of the world's poorest people actually live in pockets of deprivation within middle-income countries. In China and India hundreds of millions of people live below the international poverty line of $2 per day – half the price of a cup of coffee in Australia. In many middle-income countries basic health and education services are weak and unreliable. Here people's chief aspirations are for the health, wealth and freedom that development brings.

These two global challenges – poverty and the environment – are the twin imperatives of the twenty-first century. One ravages billions of people alive today; the other threatens billions yet unborn.

Those who are focused mainly on the environmental challenge are usually based in rich countries. In these countries the "green" agenda is to reduce our energy consumption, raise the price of fossil fuels, reduce the impact of mining, scale back our land use, practise sustainability, cover fields with wind and solar power generators, return to organic farming and preserve ancient forests. But green groups miss the point that many of these solutions don't work for the poor. The developing countries want more economic growth, more food for their hungry people, more light in their dark villages and more vehicles shipping goods from farms to markets.

For two decades rich countries have tried to force poor countries to accept their solutions to climate change, including a binding treaty to cut global emissions and a global pricing scheme to raise the cost of fossil fuels. Across the negotiating table in Copenhagen, developing countries rejected this approach. The lesson of Copenhagen is that rich countries can no longer impose solutions to global problems that ignore poor countries or assume their acquiescence in the rich countries' agenda. We now need new solutions to climate change and other environmental challenges that work for rich and poor alike.

In what follows I will apply the lesson of Copenhagen to two of the biggest environmental and economic challenges we face: resource scarcity and climate change. In both cases we find ourselves in a stalemate. In both cases we need new frameworks that reconcile progress and planet by harnessing technological means to achieve green ends.

Jeremy Grantham is one of the world's most successful businessmen. Grantham's investment company manages a staggering $130 billion on behalf of clients around the world – that's nearly the size of New Zealand's economy invested through one firm. People trust Grantham with their money because he has a rare knack for picking investment trends. He predicted Japan's crash in the late 1980s and then steered his clients away from technology stocks before the dotcom bust of the late 1990s. In 2007 he famously predicted the US housing crisis and the credit crunch that followed. Grantham's investment calls fill financial textbooks.

In April 2011 Grantham sent a letter to his investors entitled "Time to Wake Up." He warned his clients that the world is running out of resources. Our reserves of metals, hydrocarbons, land and water cannot keep up with the insatiable demand from developing countries, particularly China and India. We are, Grantham says, facing a "paradigm shift" that will rival the Industrial Revolution for its impact on our economy. The recent spikes in the price of gold, copper, coal, iron ore and oil are evidence that "we now live in a different, more constrained, world in which prices of raw materials will rise and shortages will be common."

What makes Grantham's predictions so startling is that they are so out of character. Grantham is a famously shy, softly spoken Englishman. He has built his extraordinary reputation on his ability to see through hype, by keeping a cool head through bull markets and riding out periods of irrational pessimism. But this time, Grantham says, it's different: "From now on, price pressure and shortages of resources will be a permanent feature of our lives ... We all need to develop serious resource plans, particularly energy policies. There is little time to waste."

Grantham is not alone. Concern is mounting that ever-greater human consumption of resources cannot be squared with a finite planet. "We talk scarcenomics," says Craig Cogut, the founder of Pegasus Capital Advisors. "When you look at food and water and energy, you see rising populations and resources becoming scarcer." People used to worry about running out of oil, but many environmentalists now say we're running out of planet.

## Food scarcity: can we feed another 2 billion?

The number of people on our planet is increasing at an alarming rate. To get a sense of how fast, consider the following time-line. It took nearly 10,000 years, after the birth of civilisation in the Fertile Crescent, for the world's population to reach 1 billion in 1804. From there it took just 123 years for the world's population to reach 2 billion in around 1927. The third billion came with breathtaking speed just thirty-three years later in 1960. The fourth billion took only fourteen years. The fifth billion took thirteen years. In the last two decades we have been adding people to the planet at the rate of a billion every twelve years.

The global population is set to grow by another 30 per cent by mid-century from 7 billion to 9.3 billion. Experts are asking whether the planet can produce enough food to sustain this increase. With nearly a billion hungry people on the planet already, the question is how we can possibly harvest enough food to support another 2.3 billion mouths – the equivalent of adding two extra Indias to the global family.

Not only will we have to feed more people, but rising living standards also mean that each person is eating more. As the billions living in China, India and other emerging countries ascend out of poverty, they are demanding more and better food. Already meat consumption in China has more than doubled since 1980. This is increasing the strain on our agricultural systems because every kilogram of meat served to a human requires seven kilograms of grain to produce.

The combination of rising numbers and rising appetites means that grain production will need to nearly double by mid-century. In the past, food production could be increased simply by bringing new land into cultivation. But as cities encroach onto farmland and populations grow, the land available for agriculture dwindles. In 1900 each and every human had 8 hectares of land to sustain them; today the number is 1.63 and falling. Producing twice as much food on less land will be a monumental task. The Australian science writer Julian Cribb goes so far as to say, "The central issue in the human destiny in the coming half century is not climate change or the global financial crisis. It is whether humanity can achieve and sustain such an enormous harvest."

Yet environmentalists warn that, even if it is possible, our planet won't survive the doubling of global agriculture. They argue that modern agricultural practices are unsustainable, using soil and water much faster than they can be replenished, thereby reducing the capacity to grow crops in the future. About half of the woodlands that once covered the earth are gone. Each year another 130,000 square kilometres of tropical forest disappear, most being cleared for agriculture.

Modern agriculture is wreaking havoc on natural ecosystems. Nitrogen-based fertilisers are altering the biology of water systems, eutrophying rivers and lakes. Pesticides are reducing biodiversity in soil structures. Grazing livestock are responsible for millions of tonnes of methane emissions. If we continue on our current path, environmentalists say, we will deplete our aquifers, destroy remaining rainforests, gut the oceans and fill the atmosphere with greenhouse gases.

What we face is a historic double impact: unprecedented growth (in human numbers and appetites) coincident with unprecedented constraints (on land availability, soil quality, water, fuel and fertilisers). How can we accommodate more people, more production and more consumption on the same planet?

*

Faced with this dilemma, it is worth looking over our shoulder. Our current predicament is not the first time in history that human population growth has seemed at odds with the health of the planet.

In 1798 Thomas Malthus made perhaps the most famous doomsday prediction in scientific history. Malthus believed humans would face an inescapable conflict between infinite human progress and finite natural resources. Population, he forecast, would grow without bounds, but food production would be limited to the yield of available land. From this inexorable calculus Malthus predicted that the ever-expanding human population must inevitably run out of food. The world would be struck by "sickly seasons, epidemics, pestilence and plague" until "premature death must in some shape or other visit the human race."

Malthus' prediction has cast a shadow over human progress for two centuries, furnishing scholars of history with a haunting premonition that human success would also carry within it the seeds of our own destruction. "The power of population," Malthus said, "is indefinitely greater than the power in the earth to produce subsistence for man." In conquering the earth, we would destroy ourselves.

Several times during the nineteenth century, the world veered close to the Malthusian precipice. Food shortages, violence and typhoid epidemics took hundreds of thousands of lives across Europe during the "Year without Summer" of 1816. By mid-century the Potato Famine drove a million Irish to North America and a million more to their graves. "From the cabins and the ditches, in their charred, uncoffin'd masses," wept the nationalist poet Jane Wilde, mother of Oscar. On the other side of the world, intermittent hunger plagued the later Qing Dynasty, with a series of famines between 1810 and 1849 killing countless millions.

Famines and pandemics were all too frequent, culling many and stalling human progress. But little by little, events conspired to give humanity an advantage. One such lucky event occurred in the second half of the nineteenth century, when a chance discovery by a European explorer delivered a vital boost to agriculture. Alexander von Humboldt's samples of Peruvian guano, brought home in 1804 from his voyage to South America, amazed European farmers. The product was formed from centuries of accumulated seabird droppings on the rocky coast and islands of Peru. Rich in phosphorus and nitrogen, guano had a magical effect on crops. Fertilised with it, European fields yielded more wheat, barley, oats and vegetables than ever before. Famines were rarer, populations grew and guano became one of the most highly prized resources in the world. For a time it seemed that this humble excrement had cut European civilisation free from its Malthusian moorings. Guano supported a massive rise in agricultural productivity and helped Europe's population to double by the end of the century.

However, not for the last time in human history, this non-renewable resource was imprudently managed. From 1840 to 1880 more than 20 million tonnes of guano were mined and shipped from South America around Cape Horn to Europe. Soon the deposits had been mined away. In 1898 the president of the British Association, William Crookes, warned that supplies were nearing exhaustion, because "for years past we have been spending fixed nitrogen at a culpably extravagant rate." Europe now found itself in a precarious position. Its booming population was supported by flourishing farms producing higher crop yields than ever before, but without fertiliser these yields would fall and famine would follow. Crookes warned that unless a substitute source of nitrogen could be found, England would be in "deadly peril of not having enough to eat." Suddenly, Malthusian pessimism was back.

Once again, in the teeth of crisis, providence smiled. The looming fertiliser disaster was averted less than ten years later when a German chemist used steam and methane to harvest the abundant nitrogen in the atmosphere. Fritz Haber was able to produce fertiliser almost literally from thin air. When the Swedish Academy of Sciences awarded Haber a Nobel Prize for Chemistry in 1919, it credited him with "an exceedingly important means of improving the standards of agriculture and the well-being of mankind." This was no exaggeration. Indeed, Fritz Haber's work has a credible claim to being the most important scientific discovery of the twentieth century. Today the Haber process produces around a million tonnes of nitrogen fertiliser every year and supports the production of billions of tonnes of food. It is said that 2 billion people on our planet would not be alive without it.

Von Humboldt's application of guano and Haber's chemistry were just two in a long series of scientific innovations that have pushed back the

Malthusian constraint and advanced human progress. Further discoveries in the twentieth century – including Norman Borlaug and Orville Vogel's high-yield wheat varieties, and improvements in irrigation, pesticides and fertilisers – unleashed a revolution in world food production. The cumulative impact of these innovations is that our planet in 2011 can support seven times as many people as in 1800 with the same amount of land, water, air and sun.

Why was Malthus wrong? His mistake was to miss the long-term implications of the compounding technological innovations taking place around him. As we approach new global challenges, we should remember that for centuries the essence of human success has been the ability to harness technology to break through the planetary constraints on our progress.

*

Malthus' lesson has implications for our current global agricultural challenge. Today we face an apparent conflict between progress and the environment: modern agriculture is damaging our environment, but the world needs more food. How can this dilemma be resolved?

In rich countries, the green movement urges us to give priority to the environment. It proposes a model of "sustainable agriculture" based on organic farming techniques. The organic food industry is prospering on a popular tide of affluence, health and environmental consciousness. Total farmland under organic management has nearly doubled in the last ten years and organic produce is a hit with consumers, becoming one of the fastest growing grocery categories on supermarket shelves. The goods are marketed to appeal to environmental sensibilities. Earthbound Farm, one of the major suppliers of organic produce in the United States, touts the credentials of its operations in this way: "we honor the fragile complexity of our ecosystem, the health of those who work the land, and the long-term well-being of customers who enjoy our harvest … Organic farming encourages an abundance of species living in balanced, harmonious ecosystems."

To its supporters in rich countries, organic techniques are a means to reconcile food production with environmental constraints. The problem is, organic farming isn't a long-term solution for future food production in the poor countries. Its advocates, like some of the climate activists at Copenhagen, misunderstand the needs of the poor. Organic farms produce lower yields per acre, even as the world needs more food with less environmental impact. The Nobel Prize–winning agronomist Norman Borlaug points out, "If all agriculture were organic, you would have to increase cropland area dramatically, spreading out into marginal areas and cutting down millions of acres of forests." Borlaug also says, "This shouldn't even be a debate. Even if you could use all the organic material that you have – the animal manures, the human waste, the plant residues – and get them back on the soil, you still couldn't feed more than 4 billion people."

Furthermore, the health benefits of organic food are illusory. In rich countries, organic producers have capitalised on safety concerns with clever marketing – "don't panic, buy organic!" has been one of the best lines – but beneath the slogans there is scant scientific support for their health claims. Sir John Krebs, the head of Britain's Food Standards Agency, says, "The current scientific evidence does not show that organic food is any safer or more nutritious than conventionally produced food."

Organic farming also isn't necessarily better for the environment than conventional farming. Because yields are lower, organic farming must use more land to produce the same amount of food. And organic fields are sometimes more chemical-intensive than conventional farms. Organic guidelines in many countries allow the use of "naturally occurring" chemicals, even though these can be more toxic than synthetic chemicals. For example, in the United Kingdom organic potato farmers protect against blight using naturally occurring copper sulphate compounds, which are significantly more toxic than fungicides used on non-organic potatoes. "The trouble is, organic farmers haven't got anything else to replace it," says Professor Tony Trewavas, an Edinburgh University plant scientist. "Blight destroys the whole crop – it gets into the leaves and you end up with nothing. Organic farmers cannot afford to lose a crop."

The debate over organic farming goes to the heart of how we deliver the agricultural revolution that will be required to feed humanity over the next fifty years. Green groups are right to point out that our current system of industrial agriculture has no chance of reconciling the demands of a growing population with the constraints of the planet. But their proposed solutions don't work for the poor.

In the twenty-first century, we should look once more to technology and innovation for solutions. For example, genetically modified crops can produce more food with less environmental damage. GM crops use less land and fewer chemicals than conventionally farmed crops. GM soy, canola and cotton crops have enabled farmers to stop spraying their crops with nasty chemicals (including atrazine and metribuzin, which inflict significant damage on freshwater wildlife) in favour of relatively benign glyphosate herbicides. And GM food has an extraordinary safety record. In the past fifteen years, more than 2 billion acres of GM crops have been planted and hundreds of millions of people have eaten GM produce and there is still not a single case of harm to human health. The World Health Organization reports that "no effects on human health have been shown as a result of the consumption of such foods by the general population." Indeed, genetically modified food has the capacity to make our food healthier, by creating more nutritious strains of wheat and rice that could improve the diet of billions.

Despite the promise of GM agriculture, it is fiercely resisted by environmentalists, rejected by organic farmers and has some high-profile critics. Prince Charles recently claimed that GM crops have the potential to destroy the earth's food supply. He claimed that GM food is a "gigantic experiment ... with nature and the whole of humanity which has gone seriously wrong ... and if they think it's somehow going to work because they are going to have one form of clever genetic engineering after another, then ... count me out, because that will be guaranteed to cause the biggest disaster environmentally of all time." The organic movement has success-fully prosecuted its anti-GM case, holding up safety trials and regulatory approvals in many countries. Activists in Europe and the developing world have managed to keep GM crops bound up in red tape, hindering research and slowing the uptake of such crops by farmers.

Recently, the debate over GM food in Australia turned ugly. In the early hours of the morning of 14 July 2011, members of Greenpeace broke into a CSIRO experimental station near Canberra. Activists dressed in hazard suits used whipper-snippers to raze the small field, destroying a trial crop of GM wheat. "We had no choice but to take action to bring an end to this experiment," a Greenpeace spokesperson said. "GM has never been proven safe to eat and once released in open experiments, it will contaminate. This is about the protection of our health, the protection of our environ-ment and the protection of our daily bread."

The problem here is not that Greenpeace is questioning GM food. That is its members' right. But they should not be destroying the CSIRO trials that will help us answer those questions. And they should not be rejecting a potentially valuable new technology out of hand. "The environmental movement has done more harm with its opposition to genetic engineering than with any other thing we've been wrong about," says the environmen-talist Stewart Brand. "We've starved people, hindered science, hurt the natural environment, and denied our own practitioners a crucial tool in defence of a bizarre idea of what is natural."

GM food certainly isn't the whole answer to the food challenge. We also need improved soil fertility and land management, better water and transport infrastructure in poor countries, improved warehousing and reduced waste right along the supply chain. But GM food is one promising example of how we can use technology to reconcile competing priorities. It is one more innovation besides Von Humboldt's guano discovery and Haber's nitrogen fertiliser that has the capacity to expand the realm of human possibility beyond the constraints of our environment.

## Resource scarcity: are we running out of materials?

Beyond food, there is now a broader fear that we are running out of a wide range of resources. Oil, coal, iron ore, copper, silver, palladium, nickel, gold, lead and a range of other commodities have all experienced extraordinary price rises in recent years. Unlike food, these commodities are non-renewable. Once gone, they cannot be replenished.

On current rates of growth, the global economy will be nearly fifteen times bigger than today's economy by 2100. Many experts are asking how, if our fragile planet is already stretched, we can possibly support this massive growth – and a growing group of economists believes that we cannot. "What on earth does such an economy look like? What does it run on?" asks Tim Jackson from the University of Surrey. "It's totally at odds with our scientific knowledge of the finite resource base and the fragile ecology on which we depend for survival."

Instead, some experts believe that we will soon witness the "end of economic history" – a period in which resource constraints choke off economic growth. Jackson's recent book *Prosperity without Growth* has been a global sensation in environmental economics. His thesis is simple. Economic growth is using our resources and damaging our environment at a rate that cannot be sustained. Jackson says it's delusional to think we can keep growing indefinitely without hitting the physical limits of the planet. He proposes an alternative economic model which seeks to fulfil human needs without economic growth. The Australian environmentalist Paul Gilding says we should prepare for the end of continuously expanding economies, and that "the sooner we start getting ready, the better off we'll be when it arrives."

Is this really the future we face: either a ravaged earth or the end of economic growth? Neither of these options is palatable. Continuing to gobble up resources at current rates amounts to pillaging the planet; and yet stopping economic growth would strangle the aspirations of people all over the world. The poor countries need more growth and more consumption of resources to improve their patchy infrastructure, build up their ramshackle cities, develop their industries and raise living standards.

*

Again, history provides an instructive guide to our current predicament. Thirty years ago an economist, Julian Simon, and an environmentalist, Paul Ehrlich, made a bet about the future of resource scarcity. The "Simon–Ehrlich wager" has become a famous test of two alternative visions of humanity's future. Ehrlich envisaged a future of deprivation and economic decline brought about by imminent shortages of key materials. In his bestselling 1975 book *The End of Affluence*, Ehrlich predicted, "Before 1985 mankind will enter a genuine age of scarcity." He forecast that commodities such as iron, nickel, copper and even paper would become increasingly difficult to obtain and thus much more expensive. Ehrlich's conclusion was alarming: "starvation among people will be accompanied by starvation of industries for the materials they require."

The book's dire predictions tapped a mood of anxiety in the 1970s, as rises in commodity prices seemed the harbinger of crippling future shortages. A few years later, in 1977, no less a figure than the president of the United States, Jimmy Carter, warned Americans that declining resources would be "the greatest challenge our country will face during our lifetimes."

Julian Simon, an economics professor, was deeply sceptical of this pessimism. He believed technological progress would ensure that humans never ran out of key materials. Simon proposed to Ehrlich a public wager. He bet Ehrlich that any commodity he named would fall in price over any time period he specified. Ehrlich accepted, picking five metals that he thought would undergo big price increases: copper, nickel, chromium, tin and tungsten. Ten years thence, 29 September 1990, was designated as the pay-off date. If the prices of the various metals increased over ten years, Ehrlich would win the bet.

Over the decade, many people watched the Simon–Ehrlich wager closely. Initially the underlying trends appeared to support Ehrlich. The global population increased at its fastest rate in history, adding nearly a billion people to the planet over the ten years. When nickel and chromium both experienced large price spikes in the mid-1980s, many thought Simon would lose the bet. However, by the end of the wager period in September 1990, the price of the basket of metals had fallen. In fact, each metal had fallen in price, without exception. As a result, in October 1990 Paul Ehrlich mailed Julian Simon a cheque to settle the bet in Simon's favour. Simon characterised his win as follows:

> More people, and increased income, cause resources to become more scarce in the short run. Heightened scarcity causes prices to rise. The higher prices present opportunity, and prompt inventors and entrepreneurs to search for solutions. Many fail in the search, at cost to themselves. But in a free society, solutions are eventually found. And in the long run the new developments leave us better off than if the problems had not arisen. That is, prices eventually become lower than before the increased scarcity occurred.

Ehrlich was wrong for the same reason Malthus was wrong. Technological developments provide solutions to resource scarcity. By motivating entrepreneurs, periods of high prices can actually spur innovation and investment that ultimately lead to lower prices and more abundant materials.

*

The theory of "peak oil" is another example of how predictions of scarcity can often be wide of the mark if they disregard technological change. Peak-oil theorists have long claimed that the world's production of oil will soon plateau, leading to shortages, price rises and economic malaise.

Peak-oil theory is nearly as old as oil itself. For more than a century, pessimists have been predicting that the world's oil stocks would soon begin to decline. In 1914 the US government announced that production would run out in ten years. In 1939 the US Department of the Interior declared that reserves would be exhausted by 1952. When that proved to be a false alarm, the estimate was revised to 1964. In 1978 President Carter announced that all proven oil reserves would be used up by 1990. Each time the predictions have been wrong.

Despite dozens of false warnings, peak-oil theory is still alive today. In February 2011, oil markets were startled when a cache of WikiLeaks cables from the American embassy in Riyadh indicated that Saudi Arabia could be overstating its oil reserves by as much as 40 per cent. This caused *Time* magazine to run a story, "Is Peak Oil Coming Soon?" and warn of the profound global consequences of declining oil production for inflation and economic growth. Dick Smith wrote this year that "half of all history's energy has been used in the past forty years" and warned that declining oil could lead to conflict: "The risk is that in a serious energy crisis we may fall into a period of violence and lawlessness as the competition for dwindling resources increases."

Peak-oil theorists, like Thomas Malthus and Paul Ehrlich before them, make the mistake of assuming technology is static. When Dick Smith says that we have used up half of all history's energy, he is referring to the widely cited fact that of the 2 trillion barrels of economically recoverable oil, 1 trillion barrels have already been pumped. But Smith misunderstands how technology affects our estimates of oil reserves.

In fact, geologists have long believed there could be more than 10 trillion barrels of oil on the planet, but much of this is stored in a form or at a depth that makes it inaccessible. With the technology available a century ago, only 10 per cent of oil was thought to be recoverable from major reserves. More recently, improved extraction techniques have enabled oil companies to reach around 20 per cent – hence the oft-cited 2 trillion barrels of "economically recoverable" oil.

But the point that the peak-oil crowd misses is that improved technology and higher prices will increase average recovery rates. A rise to 35 or even

50 per cent recovery rates would boost the world's oil by more than the total of today's proven reserves.

Over time, improved extraction and conversion technology may enable us to access more of the oil in existing wells and also tap into huge reserves of oil trapped in the tar shales of Venezuela and the oil shale of the Rocky Mountains. Hitherto out of our reach, these may contain ten times the oil reserves of Saudi Arabia. That is why the world's foremost authority on oil use, the International Energy Agency, forecasts that oil consumption will continue at current rates for decades to come. The IEA says that serious reductions in oil use will come as a result of efforts to reduce climate change rather than exhaustion of fossil fuels.

When asked when he believed the world would run out of oil, Sheikh Yamani, the former boss of the Organization of Petroleum Exporting Countries, famously replied, "The Stone Age didn't end because the world ran out of stone." Yamani is right. Nor did the Bronze Age end because we ran out of bronze or the Iron because we ran out of iron. In each case innovation heralded a new era of human progress that facilitated a transition to new materials.

*

What do the Simon–Ehrlich wager and the story of peak oil tell us about the future of resources? Currently, demand in developing countries – particularly China – is extremely high. China consumes a staggering 47 per cent of all iron ore produced in the world, 45 per cent of global steel, 46 per cent of global coal and 38 per cent of global copper. These resources are fuelling China's extraordinary manufacturing and construction boom. Australians know better than most that this unprecedented demand has pushed prices high; and, depending on market fluctuations, they may stay high for some time. But what we are witnessing won't necessarily result in mass scarcity and economic decline. The forces identified by Julian Simon will intervene.

First, high prices will cause entrepreneurs to invent new materials that can substitute for scarce commodities. For example, copper's price has risen

500 per cent in the last ten years due to massive demand from the tele-communications and electricity industries. As the copper price has risen, telecommunications companies have begun replacing copper with fibre optics and power utilities are beginning to switch to aluminium cables.

Second, high prices will cause entrepreneurial people to develop alternative technologies, processes and techniques that require smaller quantities of expensive inputs. Over time more plastic has been used in the manufacture of household appliances and more glass and concrete in the construction of buildings.

Third, high commodity prices cause investors to increase mineral exploration and the expansion of new mines. It is true that minerals are becoming harder and more expensive to mine. Miners are digging deeper for lower-quality ore deposits in less politically stable countries. We have taken "the nice, simple, easy stuff first from Australia, we took it from the US, we went to South America" says the CEO of Glencore, one of the biggest commodity companies in the world. "Now we have to go to the more remote places." But this has been the case since the days when nuggets of alluvial gold could be found in nearby riverbeds. The minerals become harder to find, but our technology to extract improves in parallel. And as the poor countries develop, their resources become more accessible.

"Resource scarcity comes in a number of different flavours," says Nicholas Davis of the World Economic Forum, "and so will the solutions." Some resources, like fresh water, are truly scarce in many parts of the world and lack easy substitutes. Other resources, such as fossil fuels and many metals, may not be scarce geologically, but are becoming costly to extract. The Australian engineer James Bradfield Moody has pointed out in his book The Sixth Wave that some of the biggest gains in resource efficiency will come not from new technologies per se, but from innovations that reduce waste and improve the efficiency of resource use.

The world will experience periods of high and low prices for commodities; but we will never "run out" of minerals or energy in any absolute sense unless we fail to innovate. If we focus on innovation, we will gradually switch from scarce commodities to alternative materials, in the same way that fibre replaced copper in telecommunications and plastic replaced steel in household goods. If scarcity, whether temporary like the 1980s spike in nickel or permanent like the exhaustion of the supply of guano, causes prices to rise, this will only increase the incentive to innovate and invest in alternatives.

*

The question of resource scarcity is this: how can our planet sustainably grow enough food and yield enough materials to accommodate the poor countries – both their rising populations and rising living standards?

For some, the answer is that our planet can only survive if we compromise on progress. Thomas Malthus, Prince Charles, Jeremy Grantham, Tim Jackson, Paul Ehrlich, Greenpeace and many others predict imminent calamity unless we drastically reduce consumption, shrink our population, return to simpler agricultural techniques, scale back mining of the earth's finite resources and even rethink economic growth. But these solutions fail to heed the lesson of Copenhagen. Poor countries need more food for their hungry people, more infrastructure in their underdeveloped cities and more economic growth to lift billions out of poverty.

Another answer is that we should continue on our current path. Free market optimists and climate sceptics believe we should pursue progress in the hope that the planet will adapt, solutions will emerge, or the problems will turn out not to be as severe as we had feared. But this would be a fool's gambit. We cannot continue to pump greenhouse gases into the atmosphere, clear forests, degrade arable land, devastate the oceans and damage ecosystems.

The choice between progress and planet is a false one. True solutions must lie with approaches that seek to reconcile human advancement and environmental protection, often with the assistance of technological breakthroughs. Technology has been a generous benefactor throughout human history. But hindsight runs the risk of making progress appear easy: technology can seem a conveniently simple answer to the problems we have faced, a magic elixir that has helped us to have our cake and eat it too. It is important to remember that innovation is no simple task. It often arrives unpredictably, after hundreds of failures, decades of wasted effort and mountains of lost treasure.

In the next section I'll apply the lesson of Copenhagen and the promise of technology to the problem of climate change. So far the world has failed to deal with climate change in a way that is consistent with the aspirations of poor countries. This failure has hampered global efforts to cut emissions for two decades. Tackling climate change in a way that works for the poor will be first and foremost a technological challenge – and it will be one of the most difficult challenges we have ever faced.

Human consumption of both food and resources has grown at an extraordinary rate over the past two centuries. But neither has grown as fast as the consumption of humankind's greatest modern asset: energy.

The first humans were, like all animals, born into a world where their only source of energy was their own muscles. Since then, the steady shift towards more powerful energy sources has been integral to human progress. Egyptians and Romans harnessed the power of slaves to build monuments, propel ships and work the land. Europeans used animals to power agriculture in the Middle Ages, with oxen hauling carts, ploughing fields and turning mills. Later, water and wind power were tamed for milling grain, turning hammers, cutting timber, crushing ore and shaping metal. The medieval scholar Lynn White has written that:

> The chief glory of the later Middle Ages was not its cathedrals or its epics or its scholasticism: it was the building, for the first time in history, of a complex civilization which rested not on the backs of sweating slaves ... but primarily on non-human power.

The array of energy sources in the Middle Ages was certainly impressive. Chemical energy in the form of fuelwood and tallow candles generated heat and light. Mechanical energy from beasts, wind and water was applied to agriculture and manufacturing. But almost all medieval power was renewable energy – and this was the problem. All these sources of power renewed themselves too slowly. To escape the dirt and drudgery of the Middle Ages and put humanity on the path to the health, wealth and quality of life we enjoy today, a more intense source of power was needed.

The discovery of coal delivered a massive step-change to human civilisation. Coal was a powerful, compact and transportable energy source. "We may well call it black diamonds," wrote Ralph Waldo Emerson in 1860. "Every basket is power and civilization. For coal is a portable climate.

It carries the heat of the tropics to Labrador and the polar circle; and it is the means of transporting itself whithersoever it is wanted."

Coal's greatest gift to us may have been the Industrial Revolution — perhaps the most important development in economic history. Coal powered the steam engines that turned the pistons and dynamos in workshops and trains. Coal made iron into steel for factories, rail lines and machines. To generate the power sourced from coal at the height of the Industrial Revolution, Britain would have needed to burn a quantity of wood equivalent to a forest the size of Scotland, or the muscles of 40 million slaves, or the harnessed power of 6 million horses. Put simply, without coal there would not have been enough power in England to sustain the Industrial Revolution.

Today the average person uses fifty times more power than their equivalent did in 1750, does 250 times more travel and uses nearly 40,000 times more lighting. These astonishing improvements in the most basic elements of our quality of life — heat, light and mobility — were made possible by the intense energy stored in fossil fuels, which make up 85 per cent of our energy supply. Coal has fuelled humankind through its most extraordinary period of achievement.

Of course, the problem with coal is its environmental effect. President Obama's energy secretary has put the point simply: "We have lots of fossil fuel. That's really both good and bad news: We won't run out of energy, but there's enough carbon in the ground to really cook us."

The climate challenge is one of the gravest environmental threats humanity has ever faced. For two decades our solutions to climate change have focused on a global treaty to enforce emissions cuts and a global framework to increase the cost of fossil fuels. In Copenhagen both of these policies failed. Poor countries would not sign a binding treaty which may compromise their ability to find a path out of poverty. And they would not accept any scheme which increases the cost of energy for their citizens (other than notional schemes with little impact). If Copenhagen taught us anything, it is that we need a new approach.

## The Kaya Identity: are Australia's targets weak or strong?

The political debate about climate change is awash with unhelpful jargon and prejudice. Yet the essential elements of the problem are remarkably simple to outline. I will begin with a basic concept called the Kaya Identity. This is the equation at the core of the complex climate scenarios used by the Intergovernmental Panel on Climate Change (IPCC), academic modellers and environmental institutes around the world. You only need primary-school maths to grasp the Kaya Identity, and once you do, you'll have the tools to understand the practical challenge.

Just four components make up global greenhouse-gas emissions. The Kaya Identity summarises how changes in these components affect emissions. The Kaya Identity states that total emissions (E) are made up of population (P); consumption per person, measured as GDP per capita (GDP/P); energy efficiency, measured as energy (N) used per unit of GDP; and energy emissions, measured as the emissions released (E) per unit of energy (N). So the identity, written as an equation, is:

Emissions = population × consumption per person
× energy efficiency × energy emissions

or

$$E = P \times GDP/P \times N/GDP \times E/N$$

Expressing this as changes over time gives a very useful equation:

Emissions reduction target = population growth impact + economic growth impact + energy efficiency impact + energy emissions impact

So for any given growth in the economy or population, the Kaya Identity tells us how much we will need to clean up our economy (through both improvements to efficiency and shifts to clean energy) to achieve our target.

Now let's use this equation to cut through the political rhetoric. We'll begin by uncovering what Australia's emissions target really means in practice.

Australia aims to reduce its carbon emissions by 5 per cent below 2000 levels by 2020. This target has bipartisan support, is an international commitment enshrined in the Copenhagen Accord and is the primary objective of the proposed emissions trading scheme.

Australia's 5 per cent cut in emissions has been roundly criticised for being too weak. When Kevin Rudd announced the target at the Press Club in Canberra in December 2008, there were cries of "shame" from green demonstrators in the room. Don Henry of the Australian Conservation Foundation said, "The weak targets announced today will damage Australia's international reputation." Around the world Australia's target was unflatteringly compared with the European target of a 30 per cent cut, the US target of 17 per cent and the Japanese target of 25 per cent.

But is Australia's target really as weak as green groups have suggested? Let's use the Kaya Identity to dig a little deeper.

For Australia, the elements of the Kaya Identity are as follows: our emissions target is a 5 per cent cut; population growth is forecast to be 1.7 per cent per year between now and 2020; annual economic growth per person will be around 1.6 per cent per year; and energy efficiency can be assumed for this analysis to continue to improve at its historical rate of about 1 per cent per year.

Putting these four elements into the Kaya Identity gives us a value for the fifth element: how much we will need to clean up our energy supply to reach our target. This last element, "energy emissions," measures the greenhouse gases released per unit of energy produced in Australia; in simple terms it captures how clean our energy supply is. If a large share of our energy is generated from coal and other dirty fossil fuels, then our energy emissions will be high; and if we switch to cleaner energy sources, our energy emissions will fall.

The chart below shows how much we will need to clean up our emissions to reach our 5 per cent target. The point that the critics of Australia's target miss is that rapid population and economic growth are working against us. This means that relative to 2000 levels, a 5 per cent reduction in 2020 is a big ask for our economy. As of 2010, Australia's emissions were already about 5 per cent higher than in 2000. Population growth and economic growth make our target harder to achieve by 16 per cent and 15 per cent respectively between now and 2020. To overcome these headwinds, the Kaya Identity tells us that Australia will need a massive shift to clean energy if we are to reduce our energy emissions, equivalent to a 31 per cent reduction in Australia's emissions. This is an enormous challenge.

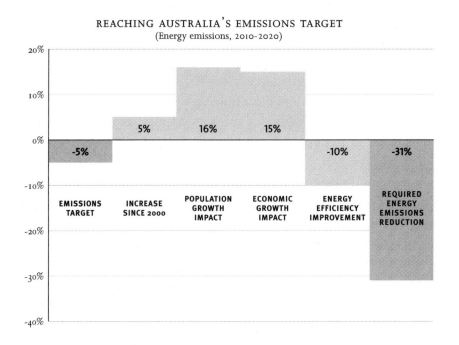

REACHING AUSTRALIA'S EMISSIONS TARGET
(Energy emissions, 2010-2020)

Australia's target may appear weak to some, but it is actually extremely ambitious. The real question is whether such a massive shift to clean energy is achievable. Energy-wise, Australia is one of the dirtiest of all the advanced economies – we have the highest emissions relative to the size of our economy. So what will it mean to clean up our energy supply by enough to produce a 31 per cent reduction in emissions?

To get a sense of the task, consider Belgium. Belgium's emissions from energy are around 30 per cent lower than Australia's. To reach our target Australia will have to clean up its economy such that we reach the same level of emissions from energy in 2020 that Belgium has reached today. That might sound simple enough, but Belgium is a highly efficient energy producer. Belgium sources more than 50 per cent of its electricity from seven nuclear power reactors (producing almost no emissions), imports a significant amount of electricity from its European neighbours (these emissions are not included in Belgium's total) and generates only 10 per cent of its electricity from coal (the dirtiest energy source).

By contrast, Australia imports no electricity, has no nuclear power and generates 76 per cent of its electricity from coal. Can Australia match Belgium in under ten years?

To do so will require a massive shift to clean power. I'll spare you the maths here (it's in the endnotes if you're interested) and skip to the summary of the task. To achieve our target, more than a third of Australia's total electricity generation will need to be clean energy by 2020. Since Australia's electricity generation will be around 300 TWH (terawatt hours) in 2020, this means we will need to build more than 100 TWH of clean-energy facilities.

This number is Australia's emissions reduction target in a nutshell. If we can generate 100 TWH of zero-emissions power, we'll catch up with Belgium. This is what our 5 per cent target means in practice.

Let's look at how hard it will be to generate 100 TWH of zero-emissions energy by 2020. Currently we generate about 9 TWH from the main renewable energy options, including 5 TWH from wind, 0.5 from solar and 3 from biofuels. To meet Australia's target, our challenge is to multiply this tenfold before 2020. Is this realistic?

First, consider the possibility of rapidly expanding our use of solar power. The Greens' leader, Bob Brown, likes to make the point that Australia's "advantage of copious sunshine is matched by the world's best technology in photovoltaics or solar panels." This is true. Australia is unquestionably well positioned to take advantage of developing solar-power technology. But can investment in solar power deliver by 2020?

On 18 June 2011, the Australian government announced it would subsidise construction of the Moree Solar Farm photovoltaic project. This project will comprise 650,000 photovoltaic panels, producing 150 megawatts of power. It will be the largest such power station in Australia, and indeed one of the largest in the world. The government will provide more than $300 million to support it.

The Moree Solar Farm is an impressive project, but it also demonstrates the wide gap between rhetoric and reality when it comes to clean energy. The project will take four years to complete, making it fully operational by the end of 2015. At that time it will deliver around 0.4 TWH of energy to the Australian grid. To reach 100 TWH of solar energy solely through projects like this, the government would need to open a project of similar size each and every week thereafter until 2020 – a total of 250 solar farms. Unfortunately, the government has provided funding for just four plants as part of its Solar Flagship program.

Second, consider wind power. In 2010 Australia produced roughly 5 TWH of wind power. To reach 100 TWH of wind power by 2020 we would have to install twenty times our current total wind-power capacity in the next ten years. This would mean building five wind turbines every single day between today and 2020. Even then, wind power remains limited by intermittency to between 20 and 30 per cent capacity. Neither wind nor solar power can currently provide continuous base-load power. Unless better storage technology is developed, these sources deliver nothing to the grid when the wind isn't blowing and the sun isn't shining.

Nuclear is another possibility. Australia could generate 100 TWH of zero-emissions energy from around twenty standard 750-megawatt reactors. But this is probably not a feasible option for Australia, at least in the short term. Nuclear power plants are expensive, require deep technical expertise, take decades to build and are opposed by both major political parties. Australia also has hydroelectric power generation – the Snowy Scheme was a world-leading project. But hydroelectricity isn't scalable because the number of rivers that can be dammed is limited and many of the best sites have already been exploited.

These scenarios, of course, illustrate the magnitude of the challenge if it is met with one source of power alone. But even combining all known technologies, Australia will not be able to meet its 2020 target of a 5 per cent cut through clean energy. In fact, trying to roll out large-scale renewable energy with current technology would be a terrible waste of money. We would spend billions of dollars installing expensive and inefficient renewable power with technology that will soon be outdated.

So far we've seen that, far from being too weak, Australia's emissions target is extremely ambitious. And, as the Kaya Identity demonstrates, the target is all but unachievable through domestic efforts.

How, then, is the government proposing to make a 5 per cent cut by 2020? The answer is that it isn't. On the government's forecasts, Australia's domestic emissions will rise to 12 per cent above 2000 levels by 2020, even with a carbon price. We will reach our target by buying emissions reductions from other countries.

The Treasury documentation states, "In the core policy scenario, the carbon price produces: around 58 Mt $CO_2$-e of domestic abatement and 94 Mt $CO_2$-e of international abatement in 2020." That is to say, two-thirds of the "carbon pollution reduction" achieved by the Australian emissions trading scheme will occur in developing countries, from whom we will buy carbon credits.

Australia's emissions reduction target of 5 per cent is not weak. In fact, it would be so difficult to achieve with domestic action alone that the government's scenarios don't even contemplate it. But to say that our emissions target is ambitious is not the same as saying it is sufficient. In fact, if Australia is to play its part in the longer term, we will need to reduce our emissions by as much as 80 per cent by 2050. As must now be clear, this is an extraordinary challenge.

## Developing countries: can we ask the poor to reduce emissions?

If cutting emissions in a rich country like Australia seems hard, spare a thought for the poor countries. Many climate campaigners in rich countries hold to the conventional wisdom that we use too much energy and need to cut down. Earth Hour, an environmental campaign that began in Australia, urges us to "unplug."

But these climate campaigns miss the most fundamental element of the global energy/climate equation: the world needs much more energy, not less. Developing countries desperately need more power. Nearly 1.5 billion people (or 20 per cent of the global population) live without access to electricity. For these people, every hour is Earth Hour and they don't like it.

In some African countries, electricity consumption per capita is as little as 50 kilowatt-hours per year, compared with an average of 8600 kilowatt-hours in the rich countries. Nearly 3 billion people still cook on indoor stoves using wood, crop waste or dung because they don't have access to modern electricity. Smoke from indoor stoves and kerosene lighting is one of the world's biggest killers, taking 1.5 million lives every year – twice as many as malaria.

About 400 million people in India are not connected to the electricity grid. The power that they have is produced by local generators running on diesel fuel trucked from village to village. In Chakai Haat, a village in the eastern state of Bihar, food is cooked on stoves fuelled by animal waste and kerosene lamps provide a little light. A *New York Times* article described the impact of power on the village:

> Chakai Haat once had power at least a few hours each day, and it changed the rhythm of life. Petty thefts dropped because the village was lighted up. The government installed wells to irrigate the fields.

Rice mills opened, offering jobs. The boon did not last long. Strong rains knocked down the power lines. The rice mills closed. Darkness swathed the village once more.

Developing countries have a right to develop and emerge from poverty. To do this they need more energy to run factories and shops, grow crops and deliver goods to market. Climate change cannot be an excuse for rich countries to say: "Do as we say, not as we did." Accommodating progress in developing countries will mean the world consumes more modern energy, not less – including, at least in the short term, more fossil fuels.

Take South Africa as an example. Since the end of apartheid in 1994, the South African government has embarked on a huge program to bring power to the 40 per cent of its population living without electricity. Over the last fifteen years, more than 20 million people have been connected to the electricity grid. Meeting this huge demand has required a significant investment in new coal-fired power stations, which now supply 90 per cent of South Africa's electricity.

Earlier this year, the World Bank helped finance one such massive coal-fired power plant. The 4800-megawatt Medupi generator north of Johannesburg will be one of the world's largest and will produce 25 megatonnes of carbon emissions. The World Bank's decision to provide funding was highly controversial. Green groups lobbied to block it. "This is a massive amount of international public finance going to the dirtiest form of energy," said Christian Aid's Eliot Whittington. When the time came for the World Bank's member countries to vote on the loan to the Medupi plant, the rich countries refused to support the project. The United States and United Kingdom tried to save face by abstaining. But the developing countries voted as a bloc in favour of the Medupi loan. They voted for power. It was Copenhagen all over again: rich versus poor; progress versus planet.

The developing world needs more power now to help its economies emerge from poverty, but renewable power simply isn't ready. With millions of poor people joining the electricity grid, coal is the only hope to keep the lights on. The South African minister of finance, Pravin Gordhan, describes his country's need for coal power almost apologetically: "If there were any other way to meet our power needs as quickly or as affordably as our present circumstances demand, or on the required scale, we would prefer technologies that leave little or no carbon footprint. But we do not have that luxury if we are to meet our obligations to our people and to our broader region."

A compelling conclusion emerges from the example of South Africa and other developing countries. Climate change cannot be solved by reducing energy use or making dirty fossil-fuel power more expensive. In developing countries – where most of the world's population live – there is an urgent need for the opposite: more, and cheaper, energy. This conclusion, as we will see, has fundamental implications for global climate policy.

## Clean energy: are renewables ready?

The Medupi plant symbolises the gap between need and aspiration when it comes to renewable power in poor countries. In Australia the difficulty of achieving even a modest emissions reduction target using clean energy highlights the limitations of renewable power even in rich countries.

The same is true around the world: renewable energy technology is not currently good enough to tackle climate change at a reasonable cost. Saul Griffith, founder of Squid Labs and recipient of the MacArthur genius grant, has provided a powerful illustration of the limits of renewable power. Griffith estimates that if we are to tackle global climate change, we will need to be annually generating 110,000 TWH of clean energy in twenty-five years' time (this equates to adding 13 terawatts of clean-energy power generation). Griffith has done the maths to work out what 13 terawatts of clean power looks like, using current technology. Griffith says it would mean:

- 2 terawatts of solar photovoltaic power, which Griffith estimates would require installing 100 square metres of solar cells every second for the next twenty-five years or 30,000 square miles of photovoltaic cells, and
- 2 terawatts of solar thermal, requiring 50 square metres of reflective mirrors installed every second for the next twenty-five years or 15,000 square miles of mirrors, and
- 2 terawatts of wind power, requiring twelve wind turbines installed every hour for the next twenty-five years or 10,000 square miles of wind turbines, and
- 2 terawatts of biofuels, meaning four Olympic-sized swimming pools of genetically engineered algae installed every second for the next twenty-five years or well over 500,000 square miles of pools, and
- 2 terawatts of geothermal power, meaning three steam turbines every day for twenty-five years, and
- 3 additional terawatts of nuclear power, requiring three 1000-megawatt nuclear plants built every week for twenty-five years.

Each of these tasks seems impossible, but let's be clear that Griffith isn't saying the world needs to do one of them. He is saying we need to do all of them: solar, wind, biofuels, geothermal and nuclear. The area of land required to house these generation facilities would be bigger than New South Wales. Add to this the space required for transmission lines, energy storage and support infrastructure and Griffith says you'd need to occupy the entire land mass of a medium-sized country and cover every square mile with renewable energy infrastructure. Griffith has a name for this fictional country: he calls it "Renewistan."

The Renewistan concept is an illustration of the enormous scale of the clean-energy challenge. Griffith's analysis is supported by a wide range of experts who have pointed out the inadequacy of current technology to tackle climate change. Bill Gates said recently that renewable energy "economics are so, so far from making sense." David Mackay, the chief scientific adviser to the UK Department of Energy and Climate Change, explained to the British parliament that to provide 100 per cent clean energy for the United Kingdom, "we would have to build wind farms with an area equal to the area of Wales; we would have to build fifty [nuclear power stations]; and we would need solar power stations in deserts covering an area twice the size of Greater London." Mackay believes that the political rhetoric about clean energy is lulling people into a false sense of security. In his blistering testimony to British politicians, he said, "there's a lack of numeracy in the public discussion of energy." Mackay worries that many people don't realise the true scale of the clean-energy challenge. They are getting the message that tackling climate change will mean a few more wind farms and solar water-heaters whereas the reality is that it will require a mammoth transformation of the way we produce and consume energy. Saul Griffith says that building the clean-energy infrastructure to address climate change will be one of the biggest engineering and technological feats ever attempted. "It's not a Manhattan project or Apollo," says Griffith. "This is much more like World War II."

In Australia, the Greens have proposed that our energy should come from 100 per cent renewable power "within decades." The Greens imply that the only barrier to renewable energy is political will and policy planning. "Right now, Australia's development of renewable energy electricity is uncoordinated and directionless," sings a typical press release. If only the money and planning were in place, we'd be on track. This is wrong. The missing element is technology, not political will. Renewable energy technology simply isn't good enough to meet our 5 per cent emissions reduction target at a reasonable cost. When the Greens imply that 100 per cent renewable power within decades is affordable and achievable in the near term, this is misleading to those of their supporters interested in practical solutions to climate change.

*

The Renewistan concept highlights two key elements of the clean-energy challenge. First, our current technology is not good enough. Some technologies are closer than others to being commercial, but unfortunately those that are closer, such as hydroelectricity and onshore wind power, are the least easy to scale up. Existing technologies will, of course, improve over time, but the challenge is such that we will need major technological breakthroughs before Renewistan is affordable and achievable.

The second element of the clean-energy challenge highlighted by Renewistan is that renewables will not, by themselves, be enough. We will need to mobilise every clean-energy technology we have. In some countries that will include nuclear. And it will include clean coal.

Nuclear power has been flatly rejected by green groups, which want to achieve emissions cuts using renewable energy alone. In Australia the anti-nuclear movement has deep roots and broad support. The Movement Against Uranium Mining and the Campaign Against Nuclear Energy, both formed in the 1970s, morphed into a broader environmental opposition to nuclear energy fronted by Friends of the Earth and the Australian Conservation Foundation. I am in sympathy with the original ideals of the anti-nuclear movement. In the 1970s, the major global risk was a war between superpowers, rather than global warming, so opposition to nuclear power seemed prudent. But things change. Now opposition to nuclear power seems to misunderstand the balance of risk.

For some countries nuclear power will be an essential weapon in the fight against climate change. For one thing, nuclear is cheaper and more reliable than renewable energy. A single 1000-megawatt reactor produces the same power as 770 square kilometres of wind turbines. Nuclear power can also be scaled up. France added 48 gigawatts of nuclear capacity – roughly equivalent to the entire capacity of Australia's electricity system – in just over a decade. France now produces nearly 80 per cent of its power through nuclear reactors and has among the lowest emissions in the industrialised world.

Nuclear power is, on many criteria, also better for the environment than currently available renewable technology. Massive volumes of concrete, steel, glass and rare elements will need to be mined and manufactured to produce solar panels and wind turbines if renewable energy facilities are rolled out at scale. Vast natural areas would also need to be used as locations for solar facilities and wind farms. For this reason some members of the green movement are starting to question the environmental costs of such projects. "As a 'Green' I care intensely about land-sparing, about leaving land for nature," says Jesse Ausubel, the director of the Program for the Human Environment at Rockefeller University. He argues that many green groups don't appreciate what extensive renewable use will mean: "Nuclear energy is green. Renewables are not green."

Environmental groups would also be wise to drop their objection to clean coal (an umbrella term to describe a range of proposed techniques to reduce emissions from coal-fired power stations). Many campaigners are quick to dismiss the promise of clean coal as an alibi for the coal industry to roll out new plants. "There is no such thing as clean coal and there never will be. It's an oxymoron," says the anti-coal Sierra Club. They have a point. Developments in clean-coal technology – in particular, proposals to pump carbon-dioxide emissions from coal-fired power plants underground (called "carbon capture and storage" or CCS) – have been extremely disappointing. But whether we like it or not, new coal-fired power plants continue to be installed. The IEA reports that coal has been the fastest-growing global energy source over the last decade, meeting 47 per cent of new electricity demand. Chinese coal consumption is increasing at a rapid rate, doubling over the five years from 2002 to 2007.

For this reason, clean-coal technology is an essential part of any realistic plan to tackle climate change. In the pathway proposed by the IEA, CCS accounts for one-fifth of the envisaged cut in emissions by 2050. This will require 100 CCS projects to be operational by 2020 and 3000 projects by 2050. The IEA acknowledges that this is ambitious, but says it's essential. The IEA concludes that without clean coal, the overall cost of tackling climate change increases by 70 per cent.

Nuclear power and clean coal both have drawbacks, but what is the alternative in our present situation? Green campaigners who insist that deep near-term emissions cuts can be achieved with a mix of conservation, wind and solar are wrong. They haven't done the maths. When green groups criticised Saul Griffith for his support for nuclear power, his response was casually scathing: "I'm not trying to be pro-nuclear, I'm just pro-arithmetic."

For over forty years, some green activists have unwittingly made the climate problem worse through their opposition to hydroelectric and nuclear power. The consequence, intended or not, was that coal-fired power plants were built instead. More and more members of environmentalists groups now recognise this has been a terrible blunder. Mark Lynas, once an anti-nuclear campaigner, admits that by stopping planned nuclear plants, he "helped release tens of billions of tonnes of carbon dioxide into the atmosphere." But many other environmentalists are refusing to change their position. By continuing to whip up public fear over nuclear power's dangers, they are contributing to the continued roll-out of coal-fired power.

Earlier, we saw that green opposition to GM crops is hindering scientific research into new food technologies. One of the fathers of climate science, NASA's James Hansen, believes that the same thing is happening in relation to climate change. "One of the greatest dangers the world faces," Hansen said, "is the possibility that a vocal minority of antinuclear activists could prevent the phase-out of coal emissions."

## Green growth: can we have our cake and eat it too?

Green groups respond to questions about the costs and practicality of clean energy with glib rhetoric about the economic benefits of a "low-carbon" economy. Tackling climate change, they say, will generate "green growth" and create "green jobs."

The Southern Cross Climate Coalition, an association of greens, unions and social organisations, has claimed that tackling climate change will lead to a jobs boom:

Australia can create up to a million clean energy jobs in at least six industries, but there's no time to waste, we must get started now. We need strong policies implemented as quickly as possible to drive investment so we can begin the economic and environmental transformation. Our international competitors are already down the track. Australia cannot finish first by starting last.

Just because something creates jobs, it doesn't follow that it is good policy. Delivering the mail eight times a day or painting Uluru purple would mean numerous jobs, but these are not good policy ideas. Job creation is only valuable if it takes workers into an industry where they are more productive than they would otherwise have been.

Proponents of "green growth" rarely define the term clearly, but to the extent that they are suggesting that tackling climate change will be good for economic growth, they are talking nonsense. Mothballing cheap fossil-fuel power plants and replacing them with more expensive renewable energy is by definition a negative productivity shock. It is true that the construction of alternative energy facilities to replace coal generators will create economic activity, but this is the same kind of economic activity that a cyclone creates when it flattens a city. People are employed in the rebuilding process, but everyone is poorer for the experience.

The truth is that acting on climate change will hit our economy in several ways. First, retiring coal-fired power stations before the end of their useful life and replacing them with clean-power infrastructure will be expensive. Second, renewable power costs more than coal, and higher power bills will suck wealth from households and businesses. Wind energy is around 50 per cent more expensive than new gas-powered electricity; solar thermal and solar photovoltaic are respectively five times and three times more expensive than gas; and geothermal and biomass generation are around twice as expensive as new gas-power generation. Ultimately these costs are borne by consumers through higher power bills or higher taxes. At current low levels of penetration, the extra cost of renewable power isn't onerous, but a large increase in renewable power would rapidly lead to sharp increases in energy costs.

There are long-term benefits associated with technological change, which will offset some of the future costs over the long term. But in the near term, it's worth being clear about the costs of addressing climate change – not to support the case for inaction, but rather to build a plan on a firm foundation of facts. Pretending that addressing the problem will be without cost is a recipe for failure. The stronger argument is to focus on the potentially immense costs of not acting.

We don't have a blank cheque. Political and economic constraints exist in both rich and poor countries. Roger Pielke Jr describes the political reality in his book The Climate Fix: "If there is an iron law of climate policy, it is that when policies focused on economic growth confront policies focused on emissions reductions, it is economic growth that will win out every time."

I saw this law at first-hand at the G8 summit in Hokkaido, Japan, in 2008. Climate change was supposed to be a major theme of the summit. A full session on climate issues was scheduled and the organisers hoped for a strong statement from the leaders that would add to the momentum in the lead-up to the Copenhagen conference the following year. But in the months leading up to the summit, something happened on world oil markets that hijacked the agenda. In early 2008 oil hit a record price of $145 a barrel, sparking a political firestorm. In Australia, for instance, petrol prices rose to $1.80 per litre at the bowser, causing an outcry from motorists and sending politicians running for cover. The Liberal Opposition immediately proposed a cut in petrol tax and the government announced the doomed FuelWatch scheme. Thousands of protesting truckers blocked roads in several European countries. Riots and demonstrations broke out in India, Sri Lanka and Malaysia.

By the time of the G8 summit, the leaders were far more interested in reducing oil prices than in increasing them. When the final communiqué emerged, politics trumped environment. Near the top, it declared that fuel prices were too high: "We have strong concerns about the sharp rise in oil prices, which poses risks to the global economy." Then, way down in paragraph 33, was a pledge to promote "market mechanisms, such as emissions-trading within and between countries … fees or taxes [to] provide pricing signals" that would raise fuel prices. The most powerful leaders in the world weren't just incoherent on the question of the price of fossil fuels: they were utterly contradictory.

London's *Independent* newspaper vividly described the contradiction: "The governments of the world say they want drastically to cut their use of fossil fuels, yet at the same time they are enthusiastically digging up any fossil fuels they can find, and hunting for more. They are holding a fire extinguisher in one hand and a flame-thrower in the other."

In rich countries, policies to cut emissions that come at a large cost to households simply will not succeed. We may wish it were otherwise, but it is not. There are too many examples that prove the point. In 2009, as President Obama was preparing to introduce his ill-fated emissions trading scheme in the United States, the Senate voted on a bill intended to ensure that climate legislation would not increase the tax burden on US citizens. The bill passed with ninety-eight votes in favour and none against. The emissions trading legislation was never passed. Even President Obama's Secretary of Energy, Steven Chu – a passionate advocate of climate action – conceded in 2009, "What the American family does not want is to pay an increasing fraction of their budget, their precious dollars, for energy costs, both in transportation and keeping their homes warm and lit." In 2010 the Pew Center found that Americans placed global warming last when ranking the importance of twenty-one issues for the year ahead. The number one issue was the economy. Number two was jobs.

When economic issues go head to head with environmental concerns in public debate or at political summits, experience shows that the economy wins out. Climate advocates may wish that the imperative to cut emissions would trump economic concerns, but ample evidence suggests it does not.

## Two decades of failure: is it time for Plan B?

We know what needs to be done to address climate change – which makes our current stalemate all the more frustrating. To give us a good chance of avoiding the worst consequences, we must limit the increase in global temperature to 2 degrees Celsius (compared with pre-industrial levels). For that to occur, greenhouse gases in the atmosphere must be kept below about 450–550 parts per million. To do this, we must reduce global emissions by at least 50 per cent by 2050.

If we are to achieve this reduction, we will need a huge improvement in energy efficiency (around 55 per cent) and a cut in emissions from energy production of more than 60 per cent, which will necessitate a massive roll-out of renewable power, at least a doubling of nuclear power and the successful application of carbon-capture technology to coal plants. Experts will argue about which technologies will make a smaller or larger contribution and which gains are more or less certain, but there is general acceptance that we will need to push hard on all and make great strides in most.

The problem is that we are moving far too slowly. The IEA says that the current rate of progress is between two and four times slower than what is required to meet the target. Even if all the commitments that countries have made so far (many of which are not yet backed by action) are fulfilled, the IEA says we would still be "a very long way short of what is required to set the world on the path to a sustainable energy system" and it would be "all but impossible to achieve the 2°C goal."

Despite decades of warnings our emissions have continued to increase. As new power facilities are installed, more than half still use fossil fuels, locking in carbon emissions for decades to come. Financial subsidies for fossil fuels still outstrip public support for clean energy. After Copenhagen it is time to ask ourselves why our global frameworks are failing.

For decades climate campaigners have poured their efforts into publicising the mounting scientific evidence of disaster. This usefully highlights the problem but does little to solve it. In 2006 Al Gore's movie provided us with a lot of "inconvenient truths" about the climate change problem, but unhelpfully glossed over the "uncomfortable facts" that show how difficult a solution will be. But these facts explain why the world has failed to address climate change despite three decades of warnings.

First, we have to accept that most of our current climate change solutions don't work for the poor. Steep cuts in emissions are unacceptable to the poor countries because billions of impoverished people don't have enough energy. Renewable energy, as we saw in South Africa, is not yet a viable solution on a mass scale. Carbon pricing doesn't work for the poor (except at very low levels) because poor people need cheaper energy. Solutions that don't work for the poor won't work for the world, because nine of the top ten fastest-growing emitters are developing countries – collectively, these countries account for the vast bulk of growth in emissions.

Second, we must accept that attempts to bring about a legally binding global agreement to curb emissions have failed. The Kyoto Protocol, despite all the fanfare, has been symbolic rather than substantive. Its record of actual reductions is poor. The United States, having refused to ratify the treaty, has increased its emissions by 17 per cent since 1990. Canada will overshoot its Kyoto target by around 30 per cent. Even Europe, which is on track to meet its targets, has mainly done so by consuming a growing percentage of emissions-intensive goods imported from developing countries. Measured on the basis of consumption (including emissions embedded in imported goods) the EU's emissions have risen by an astonishing 47 per cent since 1990. The Kyoto complaint that European countries aren't so much reducing their emissions as off-shoring them to manufacturing plants in developing countries is well founded. The huge slowdown in the economies of Japan and Eastern Europe in the 1990s is helping these countries to meet their targets, but again the reductions are coincident to, rather than caused by, the Kyoto Protocol. David King, Britain's former chief scientist, concluded that "since 1992 [when the first talks took place], 192 nations have achieved remarkably little – despite the fact that no other single topic in the world has been given so much of policymakers' time."

Third, we must recognise that renewable energy technology is currently not good enough to deliver large short-term cuts. In Australia, the objective of transferring to 100 per cent renewable energy in the near future is a pipe dream. Globally, Renewistan would be extraordinarily expensive to instal using current technology. Clean energy still costs vastly more than energy from fossil fuels. No doubt some will deny this and others will pin their hopes on a series of circulating anecdotal reports of the "next big thing" in renewables. But the fact is that renewable energy is not well developed enough to perform at sufficient scale and reliability in many rich countries, and not cheap enough to be a viable option for poor countries.

For more than two decades our global approach to climate change has failed: we have failed to establish a globally binding treaty, we have failed to effectively bring the developing countries into a global solution, and we have failed to develop new technologies sufficient to reduce emissions rapidly.

# GLOBAL CLIMATE CHANGE: PLAN B

The inescapable conclusion is that our current approach to climate change isn't working and that we need to rethink it. We need a Plan B.

The first element of Plan B is to reassess what we are trying to achieve. The objective of climate policy needs to be turned on its head. Our goal should not be to raise the cost of energy, but to reduce it. Instead of seeking to make fossil fuels expensive, we should focus on making clean power cheap. And instead of trying to reduce energy use, we should be trying to increase it in developing countries. Our goal should be to create a world with abundant, clean and cheap energy for all. This is an objective that reconciles progress and planet. This is a challenge that can bring rich and poor countries together in a common goal. If we are to address climate change, we must turn to humanity's familiar benefactor – technological innovation – and apply it to developing better clean energy.

Second, we need a complete reversal of our relationship with poor countries. Rather than trying to force them to accept unpalatable caps on their emissions, we should be trying to help them gain access to new and greater sources of energy. Poor countries are natural allies in the clean-energy challenge because they know that fossil fuels will not be enough to meet their development needs – they are already too expensive. To bring power to their populations, these countries will need abundant and inexpensive energy alternatives. Rich and poor countries should work together to develop breakthrough technology to deliver cheaper energy for the world. Only when clean energy is more efficient and cheaper than fossil fuels will it be embraced by poor countries.

Third, we need to prepare back-up plans to help us cope with the worst eventualities, which are terrifying indeed. Uncertainty is an inescapable feature of climate forecasting. Unpredictable effects – such as the "giga-burp" of methane that may be released if the northern permafrost melts – could produce unexpected calamities. If the rate of climate change accelerates unexpectedly, we could find ourselves dealing with drowning coastlines and devastated agriculture. These may not be the most likely outcomes, but prudence requires us to prepare for them.

Most of the prominent back-up plans fall under the category of "geo-engineering," the intentional modification of the earth's climate. Geo-engineering proposals include schemes to pump sulphates high into the stratosphere to reflect sunlight (this idea is sometimes referred to as a "synthetic volcano," because there is evidence that sulphur dioxide from volcanic eruptions can have a powerful cooling effect on the planet); using ocean spray to generate artificial clouds; fostering oceanic plankton to absorb more carbon; turning carbon from agricultural waste into char-coal and burying it in the ground (biochar); and global dimming using reflective mirrors in space. The field of geoengineering needs much more work. Currently the list of ideas is long, risks are high and research is thin.

Many environmentalists dismiss geoengineering as highly dangerous at best and an excuse for inaction at worst. They have a point. Geo-engineering should not be a substitute for cutting emissions, but, given the unpredictable risks humanity faces, it may have a role to play. The Canadian academics Thomas Homer-Dixon and David Keith claim that geoengineering is "so taboo that governments have provided virtually no research money." They propose that we should begin to conduct limited and controlled real-world geoengineering experiments, so that we have some knowledge if we need to act fast. "While we should all hope that we never need to play God with the earth's climate, we must also have the best science at hand to do what might be necessary if melting polar ice leads to a far more dangerous future."

*

What government policies will be required to achieve Plan B? The conventional wisdom asserts that pricing carbon through "market mechanisms" is the best and lowest-cost way to reduce greenhouse gases. This view is backed by a tide of political rhetoric about the ability of such pricing to "unlock" investment in new technologies and deliver a "clean-energy future."

Unfortunately, both the political rhetoric and the conventional wisdom are wrong. Emissions trading schemes will find the most efficient way to reduce emissions *from existing technology*, but they are not particularly effective in bringing forward the technologies of the future.

The IEA says without exaggeration that "a global revolution is needed in ways that energy is supplied and used" if we are to reach our emissions targets. It estimates that every year between 2010 and 2050 the world will need up to 14,000 new onshore wind turbines (more than twice the current rate of installation), 215 million square metres of solar panels (more than eight times the current rate), thirty-two nuclear power plants (twice as many as have ever been installed in a single year), up to 3750 offshore wind turbines (more than thirty times the current rate), fifty-five fossil-fuel plants fitted with CCS (none have been successfully done yet at scale), as well as more geothermal, solar thermal, biomass and hydro power. To achieve this will require a phenomenal policy push by governments worldwide, including massive support for investment in development, demonstration and deployment of clean-energy technologies and the infrastructure that will enable them.

But a carbon price will not be enough to bridge the wide cost gap between clean technologies and coal. Most of the scalable clean options are so expensive that the carbon price required to encourage genuinely commercial investment would need to be in the range of $100–500 per tonne, not the $20–30 now planned. The energy consultant Andrew Dyer notes that the proposed Australian carbon price may not generate "any meaningful change in our generation portfolio mix towards renewable energy." Even the Australian Greens agree. On 1 July 2011, their deputy leader, Christine Milne, told the *Climate Spectator*: "Well, I think I'm realistic enough ... to know that to get the investment into solar, thermal and geothermal and so on, you would need a very high price, which we are unlikely to be able to negotiate in the current political climate, if you like, in Australia ... it's clear to me the carbon price alone will not drive the investment in the renewable industry sectors fast enough to get large-scale renewables up and running in the time frame we need."

In addition to direct investment in clean-energy technology, we need massively increased support for research and development of new technologies because our current technologies are not good enough to deliver clean-energy infrastructure at scale at an acceptable price for the rich countries, let alone the poor ones.

But a carbon price won't be particularly effective in encouraging new technological advances. Some might argue that schemes to raise the price of fossil fuels will encourage investment in new technologies. This argument is half right and half wrong. Emissions trading schemes and renewable energy targets do create an incentive to invest in clean energy, but there is an important distinction between roll-out investment – such as building wind and solar plants with today's technology – and investment in research and development – putting resources into new breakthrough technology. This is a crucial distinction. Since better technology is the only chance we have of effectively tackling climate change, it is essential to give primacy to policies that foster innovation rather than policies that subsidise the roll-out of existing technology.

Economists broadly identify two ways of developing better technology. The first is called "learning by doing," whereby deployment leads to efficiency improvements over time – that is, if we build a hundred wind farms, the last few will be cheaper and better than the first ones by dint of experience. In his comprehensive review of clean-energy technology, Ross Garnaut provides an example of on-the-ground learning in Spain, where solar projects have achieved falling costs over time as manufacturing and assembly techniques improve. Deployment can result in important innovations, but the essential feature of learning by doing is incrementalism – gradual improvements to known technologies.

The other method is research and development. Instead of learning how better to apply existing technology, basic research investigates new methods and materials and offers the potential to produce step-change technology. It looks for breakthroughs. This type of research is often difficult to commercialise immediately and hence does not attract significant private funding, so much of it is done through universities and research organisations such as the CSIRO.

A carbon price gives very little encouragement to basic research because price incentives are meaningless for publicly funded science (like university and CSIRO research) and aren't very effective in encouraging research by private firms into less mature technologies. But it is exactly these new technologies that we need to combat climate change. Unfortunately, many governments around the world are focusing too little on basic research and too much on deployment. Bill Gates believes that the emphasis is wrong: "We're putting 90 per cent of the subsidies in deployment … not in R&D. And so unfortunately you get technologies that, no matter how much of them you buy, there's no path to being economical." We will need massive direct public support for energy technology research, with some estimates in the order of US$10–100 billion per year in additional research funding.

Carbon pricing will not by itself lead to the infrastructure investment and technological innovation we will need to reach our long-term targets. But this is not to say that such a scheme doesn't have benefits. By increasing the cost of fossil fuels (especially coal-generated electricity), the carbon price encourages energy efficiency, as higher bills nudge consumers and businesses to reduce their power use. It also tilts the balance away from coal and towards "near commercial" technologies such as gas-fired power stations. In this way it can deliver incremental reductions in rich countries' emissions over time.

The benefits and limitations of emissions trading are well summarised by Jeffrey Sachs, Director of the Earth Institute at Columbia University:

> Economists often talk as though putting a price on carbon emissions
> – through tradable permits or a carbon tax – will be enough to
> deliver the needed reductions in those emissions. This is not true.
> Europe's carbon-trading system has not shown much capacity to

generate large-scale research nor to develop, demonstrate and deploy breakthrough technologies. A trading system might marginally influence the choices between coal and gas plants or provoke a bit more adoption of solar and wind power, but it will not lead to the necessary fundamental overhaul of energy systems.

A carbon price will help us reach the "low-hanging fruit" – the relatively cheap gains from energy efficiency, energy conservation and "near commercial" clean technologies. But meeting significant domestic targets will also require huge direct government support for new energy infrastructure and a massive increase in research funding. We will make more progress on climate change with policies that make clean energy cheaper than with policies that make fossil fuels more expensive.

*

Around the world, national climate-change policies are slowly moving towards Plan B. The focus now is less on a global treaty and carbon market and more on efforts to improve clean-energy technology.

The United States has abandoned its emissions trading legislation and is instead devoting resources to clean-energy research and technology as well as to investment in new nuclear reactors. India has implemented a carbon tax at a low price of around $1 per tonne of coal for the purpose of financing its National Clean Energy Fund. The low tax is less about making energy more expensive and more about raising money to support research and extend energy infrastructure to poor areas. Corporations should be (and are) getting involved. Google is investing in clean-energy technology: it calls the initiative "RE < C" (renewable energy cheaper than coal), which succinctly conveys the goal we should all be focused on.

China is perhaps the world's best example of this new "climate pragmatism." China's climate policies often confuse Western observers. On the one hand, China is constructing hundreds of new coal-fired power stations and belching ever-greater quantities of greenhouse gases into the atmosphere. On the other, it is investing billions in renewable energy, setting serious targets to reduce the emissions intensity of its economy and trialling a limited emissions trading scheme in some provinces. What explains the alternate green and brown hues of China's energy policy?

China's energy priorities are very different from those of the West. Its first priority is development. A Chinese IPCC contributor, Jiang Kejun, says, "The fundamental homework for the Chinese government is still to make people richer ... With such a huge population and living standards still low, we can see emissions still need to keep going." That explains China's roll-out of the coal (and nuclear) generators needed to rapidly increase the supply of power to its growing economy. China's second priority is energy security. As a large country with limited domestic resources, China is worried about the security of its energy supply. The development of renewables is a way for China, over time, to strengthen its domestic supply sources and reduce its reliance on imported fossil fuels. Third, China is increasingly concerned about the environmental impact of its development. In some Chinese cities, pollution is a daily health hazard. China's leaders want energy policies that will reduce air pollution. Fourth, China wants where possible to contribute to the global emissions reduction effort. These four objectives explain China's seemingly confusing energy policy.

In Copenhagen the West tried unsuccessfully to force China to accept binding targets. Instead the West should work harder to understand China's objectives and find solutions that meet mutual goals. China is facing, perhaps more than any country, a conflict between its economic progress and its environmental constraints. China's environment minister, Zhou Shengxian, said, "In China's thousands of years of civilisation, the conflict between humankind and nature has never been as serious as it is today ... The depletion, deterioration and exhaustion of resources and the worsening ecological environment have become bottlenecks and grave impediments to economic and social development." The West should embrace China's renewable energy policies and accept that these are motivated by energy security and smog as much as by climate change. This shouldn't be a problem. The West should cooperate with China and other developing countries to realise their shared goal of cheap and clean power.

The United Nations is also moving towards a more pragmatic approach. The failure in Copenhagen has buried, at least for now, attempts to achieve a top-down binding global agreement. This need not be a permanent stumbling block. The Copenhagen Accord, widely seen as a failure because it was not a legally binding treaty, may actually represent a step forward towards a more flexible framework that allows countries to formulate their own strategies. More recently, global negotiations have focused on developing countries, technology transfer and investment in new technology.

*

On 31 December 2012 the first commitment period of the Kyoto Protocol will come to an end. Kyoto has manifestly failed. It has failed to deliver significant emissions cuts. It has failed to attract the world's largest historical emitter (the United States) or any of the developing countries that will be the biggest future emitters. Several attempts to extend the life of the commitment period and expand the ambit of the treaty have come to nothing, so we will start 2013 by stepping backwards into a world without a binding global emissions treaty.

The end of the Kyoto Protocol should mark a point of inflexion in the global approach to climate change. International agreements with binding cuts and fixed timetables cannot gain the acceptance of the poor countries that want cheaper energy and economic growth.

We need a better understanding of the poor countries. Rich countries are islands of prosperity in an ocean of poverty. We are one billion; they are six billion. For decades we have shared too little of the fruits of our economic growth with them. We should not be surprised that they are now reluctant to help us clean up the consequences of that growth.

That doesn't mean that there is no solution, just that we need a new approach. We should take a measure of comfort from the fact that environmental challenges are not new to the human story; they are a common thread through thousands of years of history and hundreds of disparate civilisations. And this is by no means the first time that continued human progress has seemed inconsistent with environmental constraints. Time and again, we have succeeded in overcoming these constraints and gone on to create an ever-safer and more prosperous world.

More often than not, our success has come from technological innovation – the human ability to imagine a better world and to bring it into being through scientific endeavour. Technological solutions aren't an easy answer to climate change. The scale of Renewistan, the elusive quest for clean coal and the graveyard of failed renewable energy projects are all evidence of how challenging the task will be. The point is not that technology is the easiest solution; it is that technology is the only solution. Only when clean energy is as cheap and reliable as fossil-fuel energy will it be adopted at scale in the poorest countries. Only when carbon capture and storage technology is mastered will China's newly built coal-fired power stations stop belching carbon into the atmosphere. Only with better energy storage can we ease our reliance on dirty transport fuels. Our efforts to develop the technology to deal with climate change must be unremitting and ever-increasing.

In the hamlet of Aneyoshi, in Japan's Iwate Prefecture, a stone tablet protrudes four feet from the earth near a walking track on the outskirts of the village. The tablet was erected more than 500 years ago, but the text carved into the face of the slab remains clearly legible: *High dwellings are the peace and harmony of our descendants. Remember the calamity of the great tsunamis. Do not build any homes below this point.*

Hundreds of these ancient markers are dotted along the northern Japanese coastline, which for centuries has lived with the spectre of seismic activity. A seven-foot stone tablet in the city of Natori warns of a link between the earth and the sea: *If an earthquake comes, beware of tsunamis.* In the coastal city of Kesennuma another ancient marker reads: *Always be prepared for unexpected tsunamis. Choose life over your possessions and valuables.*

The earthquake of 11 March 2011, registering nine on the Richter scale, was followed by a dreadful ten-metre tsunami that killed 12,000 Japanese people.

"People had this crucial knowledge," says Japanese scholar Yotaru Hatamura, reflecting on the ancient coastal warning tablets, "but they were busy with their lives and jobs, and many forgot."

In Kesennuma, one 70-year-old resident, Tetsuko Takahashi, described how she saw people returning to their homes after the earthquake to get their valuables and stow their tatami floor mats. "They all got caught," Tetsuko recalls. She watched from her hillside home as more than 500 people were swept away.

Japan's coastal tablets are one example of the human bias toward the "near and now" – our often rational approach to events close to us, but irrational response to events that are more remote. Even with the clearest warnings, we often ignore colossal risks if they are far away in time, space, probability or understanding. The same is true of many environmental risks. The American Psychological Association released a report in 2009 on Psychology and Global Climate Change which concluded that "small probability events tend to be underestimated in decisions based on personal experience, unless they have recently occurred in which case they are vastly overestimated." In Japan, people wanted to live near the coast. They rejected information that ran counter to this lifestyle choice.

This essay has considered some of the biggest risks facing humankind: the risk we won't have enough to eat, the risk we'll run out of resources, the risk of nuclear power, the risk of climate change. Mostly I have focused on policy questions: how should we best deal with these risks? But the best advice of scientists and the best plans of economists are not enough. Executing such plans requires the will to act. Without a public consensus, we cannot move forward.

This consensus is all but absent. In Australia last year the then prime minister, Kevin Rudd, called climate change the "greatest moral challenge of our time," while the Opposition leader, Tony Abbott, described the science as "absolute crap." In Germany the debate over nuclear power is cutting the country in two. In Europe the controversy over GM food has paralysed the European Parliament for a decade. For issues that are essentially scientific, these debates have become emotionally divisive.

How can we develop the political consensus we need to tackle these challenges? You might think the answer is to provide people with more information, so that consensus can emerge from greater understanding. But you'd be wrong. Research shows that the more information we have on these issues, the more polarised people become. Polling in the United States shows that over the last decade more Democratic voters have come to accept the science of global warming. Yet over the same period Republican voters have become more sceptical. Remarkably, all the attention given to climate change over the last ten years has led to a divergence of opinion.

More facts aren't helping because when it comes to nuclear power, GM food or climate change our responses are never purely rational, fact-based assessments. There are so many facts out there that anyone can pick and choose them to align with their ideological position or justify their lifestyle. In such circumstances it should not come a surprise that scientific information is not leading to political consensus. David Ropeik from Harvard University says:

> Our perception of risk is never a neutral unbiased view of the evidence. The psychology of how we perceive and respond to risk is an affective mix of facts and how those facts feel. And once we've made up our mind about a risk, Confirmation Bias takes over and we choose to believe the evidence that agrees with what we already believe. True liberals, non-wobbly liberals, are supposed to oppose nuclear power. Period. True conservatives, for some reason, are supposed to deny climate change. Period. Find the facts that fit. Toss out the inconvenient truths. Personally denigrate anyone who disagrees, because other views are not just a different way of looking at things. They're a threat. These are litmus test issues for who gets to belong to the tribe, and who doesn't.

The American Psychological Association concluded that "there is significant variability in people's reactions to climate risks, much of which is mediated by cultural values and beliefs." Which side of these issues you find yourself on is mainly determined by your identity. Do you embrace the model of human progress or are you frightened by it? Do you trust governments and institutional authority or are you sceptical and detached?

*

In this essay I have sketched opposing perspectives on several of the challenges our planet faces. One perspective can be characterised as the "optimistic" view, another as the "romantic."

Optimists believe in the power of human progress. They believe that human ingenuity can be relied upon to overcome our global challenges and deliver steady gains in welfare. The cumulative, historical advance in scientific knowledge has conquered the animal kingdom, bent the natural world to our convenience, unpicked the darkest secrets of the universe, fought back disease and carried us through the heavens. This is the "don't worry, be happy" view of history, which trusts that technological innovation will light and smooth an infinite path of human advancement.

Optimists adhere to the Enlightenment view that history is "progressive" – a steady upward path of human advancement. Beginning in the seventeenth century, Enlightenment thinkers broke away from the Aristotelian conception of a steady natural order linking the physical and spiritual world. They criticised classical naturalism for holding back the progress of man. In his *Free Inquiry into the Vulgarly Received Notion of Nature*, Robert Boyle, one of the fathers of modern chemistry, protested that "the veneration wherewith men are imbued for what they call 'nature,' has been a discouraging impediment to the empire of man over the inferior creatures of God." Whereas classical scholars sought to live within the natural order, Enlightenment scientists aspired to conquer nature. Francis Bacon believed that technology could "establish and extend the power and dominion of the human race itself over the universe." Science, he hoped, could recreate the utopia lost when Adam was expelled from Eden.

In this tradition, optimists believe that progress is the answer to our future challenges. The free market is the best incubator of progress because it promotes the free exchange of ideas and unleashes the power of human endeavour through the mechanism of self-interest. The optimists' prescription is to increase market freedom, encourage innovation and promote human progress. Even environmental problems are best mended with economic tools. The economist Wilfred Beckerman argues that rich countries have a greater capacity to deal with environmental challenges and concludes that "in the end the best – and probably the only – way to attain a decent environment in most countries is to become rich."

Romantics have a different perspective on humankind's relationship to nature. They believe nature has an intrinsic value, distinct from its role as a support for human life. Their values emerged as an artistic and philosophical reaction to the Enlightenment's scientific rationalisation of nature. For the romantics, wilderness is powerfully sublime. And whereas the

Enlightenment scientists saw a struggle between man and nature, romantics saw a complementarity. "I wandered lonely as a cloud," wrote Wordsworth, as if man and nature could merge into a seamless continuum.

Modern environmentalists have inherited the romantic reverence for nature. According to the Gaia hypothesis of James Lovelock, our planet is itself alive – a "living earth" with the self-regulating functions of an organism. Tim Flannery, who has perhaps done more than anyone to popularise the Gaia hypothesis in Australia, believes that humans are an intrinsic element of a living earth, rather than an independent life-form on an inanimate planet. "[Humans] are a sophisticated electrochemical process that is part of earth's crust," Flannery says. "It's not like we've come from somewhere else. So life and earth are one. We are one thing." The implication of the Gaia hypothesis is that humans and nature are symbiotically connected in a delicate biological balance. Human progress, by disturbing this balance, has the potential to unleash uncontrollable forces that could overwhelm us. Lovelock says that "if we fail to take care of the Earth, it will surely take care of itself by making us no longer welcome."

Romantics believe that human progress can sometimes be a cause of, rather than a solution to, our problems. Indeed, Lovelock says, "We are in our present mess through our intelligence and inventiveness." Many environmentalists urge us to reconsider our faith in free markets and economic growth. No less an authority than our prospective head of state, Prince Charles, says that "modern progress depends inherently on the exploitation of the extraordinary bounty of nature." The Australian intellectual Clive Hamilton urges us to accept "what we have refused to admit – we cannot preserve a habitable planet and pursue endless economic growth at the same time." Dick Smith warns in his recent book, Population Crisis, that our "endless growth economy is obsolete and risky to future generations. We must plan now and begin to implement a steady-state economy." For his part, Lovelock says we need not "sustainable development ... [but] sustainable retreat."

Optimists and romantics are locked in an escalating conflict over progress and planet. The perspective each takes on the sources of human success makes a large difference to their approach to future challenges. Optimists might propose that we invest in success, and throw our weight behind human industry and science to meet the human and environmental challenges of the future. Romantics might prefer to restrain economic growth and take action to preserve what is left of the natural order rather than risking the unknown consequences of further human intervention.

\*

No one individual or group is fully hostage to the romantic or optimistic disposition. Each of us has both to a greater or lesser degree. And neither the romantic nor the optimistic perspective is more correct or more virtuous. We are perhaps at our best when these tendencies are in balance: when we recognise the problems of environmental degradation without neglecting the power of economic growth and the technological innovations that drive it.

But today we appear to be losing the balance. As the political extremes on the left and right dominate the debate on environmental issues, our ability to arrive at solutions suffers.

In Australia, for instance, issues that involve the economy and the environment have plunged our politics into one of its most tumultuous periods: mining development versus the conservation of natural wilderness; the Murray–Darling ecosystem versus the livelihoods of irrigators; the logging industry versus Tasmania's forest heritage; rising food and fuel prices versus the public concern over cost-of-living increases; business needs for skilled migrants versus urban and green anxiety over rapid population growth; and, of course, the vexed issue of climate change. The battle between romantics and optimists has become a common theme, reshaping Australian politics and breaking old allegiances.

Conflicts over economics and the environment are redrawing political boundaries: dividing country and city; driving a wedge between the left and right of the Liberal Party; separating Labor's industrial base from its urban progressive supporters; splitting the business community; contributing to the historic rise of the Greens and stirring up their counterweight in rural populism.

Few issues in Australian politics are as polarising as climate change. The right remains immutably sceptical in the face of mounting scientific evidence, preferring to speculate darkly about a UN conspiracy to take over the world. The left oscillates between Henny-Penny predictions of imminent disaster and Pollyanna assertions about a renewable energy fix just over the horizon. And all the while, global emissions grow and global negotiations remain stuck.

Bill Bryson, in his panoramic work of popular science *A Short History of Nearly Everything*, ponders two events from the 1680s: Isaac Newton's unlocking of the secrets of gravity and motion in his *Principia Mathematica* and the extinction of the dodo at the hands of Dutch settlers on the island of Madagascar:

> We don't know precisely the circumstances, or even year, attending the last moments of the last dodo, so we don't know which arrived first, a world that contained a *Principia* or one that had no dodos, but we do know that they happened at more or less the same time. You would be hard pressed, I would submit, to find a better pairing of occurrences to illustrate the divine and felonious nature of the human being – a species of organism that is capable of unpicking the deepest secrets of the heavens while at the same time pounding into extinction, for no purpose at all, a creature that never did us any harm and wasn't even remotely capable of understanding what we were doing to it as we did it.

Our "divine and felonious nature" captures the two sides of the human story. In rich countries, advances in science and technology have expanded social and economic freedom, improved health and learning, and distributed prosperity more widely than our ancestors could have imagined. But progress has its price. Each step of human advancement has left a footprint on the planet.

Today our two defining challenges are managing climate change and eliminating global poverty. In Copenhagen we learned that these challenges are inseparable. If we give priority to climate change over development, the poor countries will not agree to be part of the solution. Since they will account for the vast bulk of the growth in emissions in coming decades, nothing we do in the rich countries alone will be enough. On the other hand, if we give priority to development over climate change and continue to use resources in the way we have, we will inflict huge damage on our environment, on millions of other species, and ultimately on our own wellbeing. Nicholas Stern, the climate adviser to the British government, puts it this way:

> If we try to manage climate change by putting obstacles or appearing to be putting obstacles in the way of rising standard[s] of living in the developing world, and not just the developing world, in the next two or three decades ... if we try to manage climate change that way, we will not succeed in putting together the coalition which is vital on a global scale to manage climate change. If of course on the other hand we fail to manage climate change, then the environment will become so hostile over the course of this century and into the next that we will reverse, set back the whole story of development.

The truth is: there is no choice between progress and planet. If we focus on one, we will destroy both. The only way out of our predicament is to reconcile economic development and environmental sustainability.

In 1963 President John F. Kennedy gave the Commencement Address at American University. In the shadow of the Cuban missile crisis, which months earlier had brought the world perilously close to nuclear armageddon, Kennedy addressed the prospects for global peace:

> Our problems are man-made; therefore they can be solved by man. And man can be as big as he wants. No problem of human destiny is beyond human beings. Man's reason and spirit have often solved the seemingly unsolvable, and we believe they can do it again.

Kennedy's next utterance made it clear that his was not a blind optimism:

> I am not referring to the absolute, infinite concept of universal peace and good will of which some fantasies and fanatics dream. I do not deny the values of hopes and dreams, but we merely invite

discouragement and incredulity by making that our only and immediate goal.

Today our challenges are similarly "man-made." Kennedy's balance of optimism and romanticism is apt for our age. I hope we can summon the confidence to tackle our global environmental problems and resist false panaceas. As human activity comes to dominate the planet, both our problems and solutions are in our hands. Humanity's influence is now so pervasive that any choice we make will have a massive impact on the earth. There is no choice not to change the planet, only a choice as to how we change it. "We are as gods," says Stewart Brand, "and have to get good at it."

# NOTES

I am grateful to Saul Griffith, Rod Sims, Terrey Arcus, Nicholas Davis, Louisa Fitzgerald, Steven Kennedy and several others for useful comments and suggestions and to Chris Feik for his editorial guidance. However, the opinions expressed and any errors are my sole responsibility.

4     "Backed by thousands of journalists and green activists ...": Inevitably, the deal pursued by the rich countries wasn't strong enough to satisfy everyone, but it was certainly stronger than the deal that was delivered.

5     "a Latin American negotiator explained to me ...": This was not well enough understood by many of the journalists, negotiators and activists from rich countries. Many rich country negotiators also underestimated the residual enmity in poor countries following decades of iniquitous trade agreements and broken aid commitments. These experiences left the poor countries with a doubly distilled scepticism towards any economic pact promoted by rich countries.

5     "nearly four-fifths of the planet's ecosystems are under human influence ...": Lee Hannah et al., "A preliminary inventory of human disturbance of world ecosystems," *Ambio*, Vol. 23, No. 4/5, July 1994, pp. 246–50.

6     "the average income is less than \$2 per day ...": Another billion and a half live on less than \$2 per day in middle-income countries.

6     "Measured by weight, we make up less than half of one per cent of the animals on the planet ...": Helmut Haberl et al., "Quantifying and mapping the human appropriation of net primary production in earth's terrestrial ecosystems," *Proceedings of the National Academy of Sciences USA*, Vol. 104, No. 31, 31 July 2007.

10     "grain production will need to nearly double by mid-century ...": United Nations Energy, *Sustainable Bioenergy: A Framework for Decision Makers*, April 2007.

10-11   "In 1900 each and every human had 8 hectares ...": Julian Cribb, *The Coming Famine: The Global Food Crisis and What We Can Do to Avoid It*, CSIRO Publishing, Melbourne, 2010.

11     "another 130,000 square kilometres of tropical forest disappear ...": World Resources Institute, *Forest Ecosystems Summary Assessment*, http://archive.wri.org/page.cfm?id=183&page=content.

13     "producing higher crop yields than ever before ...": Guano was so valuable to farmers that supplies could not keep up with demand. The search for guano deposits was a major motivation for the colonisation of the Pacific with the United States occupying Midway Atoll, Jarvis Island and Baker

Island, and Germany and later Britain controlling the phosphate-rich Nauru.

13  "2 billion people on our planet would not be alive without it ...": Thomas Hager, The Alchemy of Air: A Jewish Genius, a Doomed Tycoon, and the Scientific Discovery That Fed the World but Fueled the Rise of Hitler, Harmony Books, New York, 2008.

14  "Total farmland under organic management ...": Minou Yussefi & Helga Willer (eds), The World of Organic Agriculture, Statistics and Emerging Trends 2007, International Federation of Organic Agricultural Movements, Bonn, 2007.

15  "Norman Borlaug points out ...": Borlaug was awarded the Nobel Peace Prize for his contributions to agriculture and was, until his recent death, a strong critic of organic foods.

22  "The IEA says that serious reductions in oil use ...": The IEA forecasts that global oil use will rise from 84 million barrels per day today to 99 million barrels in 2035. By that time currently producing fields will supply 20 million barrels, yet to be developed fields will supply around 45 million barrels and natural gas liquids and unconventional fuels will supply the remainder.

23  "Other resources, such as fossil fuels and many metals ...": For further information see the World Economic Forum Scenarios, http://www.weforum.org/reports/mining-metals-scenarios-2030.

23  "some of the biggest gains ...": James Bradfield Moody & Bianca Nogrady, The Sixth Wave: How to Succeed in a Resource Limited World, Vintage Books Australia, Sydney, 2010.

26  "water and wind power were tamed ...": John Langdon, Mills in the Medieval Economy: England, 1300–1540, Oxford University Press, Oxford, 2004.

26  "The chief glory of the later Middle Ages ...": Lynn White, "Technology and invention in the Middle Ages," Speculum: A Journal of Medieval Studies, Vol. 15, No. 2, 1940, pp. 141–59.

27  "the average person uses fifty times ...": Roger Fouquet & Peter Pearson, "Long run trends in energy services, 1300–2000," Centre for Environmental Policy, Imperial College London, January 2005.

28  "changes over time gives us a very useful equation ...": Decomposition is done using Logarithmic Mean Divisia Index (LMDI). See Beng Wah Ang, "The LMDI approach to decomposition analysis: a practical guide," Energy Policy, Vol. 32, 2005, pp. 1131–9.

29  "European target of a 30 per cent cut ...": Europe and Japan's targets are conditional on global action and the US commitment is a 17 per cent reduction below 2005 levels.

29  "population growth is forecast to be 1.7 per cent ...": forecasts range from around 1.5 to 1.8 per cent annual growth.

29 "annual economic growth annual economic growth per person will be around 1.6 per cent ...": The Australian Treasury estimate of real per capita GDP economic growth is 1.6 per cent per year.

29 "energy efficiency can be assumed for this analysis ...": Energy efficiency reduces naturally over time as appliances become better, cars become more fuel efficient and building standards improve. Energy efficiency improvements will be slower in rich countries than in poor countries. Hence Australia's energy efficiency improvements will be slower than global improvements. In eighteen years from 1990 to 2008 Australia was able to reduce our carbon intensity by 15 per cent. So we assume that this continues at just under 1 per cent per year. See IEA, $CO_2$ *emissions from fuel combustion – highlights*, OECD/IEA, Paris, 2010.

31 "Belgium's emissions from energy ...": It is important not to confuse emissions intensity and energy emissions. The basis of these calculations is: emissions intensity is measured as $CO_2$ emissions in kilograms and $CO_2$ per dollar of GDP is measured using purchasing power parities in US-dollar 2000 prices. Achieving Belgium's emissions intensity is roughly equivalent to achieving Australia's 5 per cent target after factoring in energy efficiency improvements. Data is from IEA, 2010.

31 "more than a third of Australia's total electricity generation ...": The main components of Australia's energy consumption are petroleum products (35 per cent used in petrol, diesel and heating fuel), coal (40 per cent primarily used in electricity generation), and other (gas around 20 per cent and renewables around 5 per cent of electricity generation). It will be difficult to shift from petroleum both because petrol is excluded from the carbon tax and technology to replace liquid fuels is in its infancy. So the shift will primarily fall on the electricity-generating sector, which is made up of 76 per cent coal, 15 per cent gas, 5 per cent hydro, and 3 per cent wind and solar. This analysis assumes we shift 25 per cent of our total energy to clean power through electricity alone, which implies a 33 per cent shift to clean-electricity production. Treasury estimates that the share of gas production will not rise materially until after 2025. See Australian Treasury, *Modelling a Carbon Price*, Commonwealth of Australia, Canberra, 21 September 2011, Chapters 4 and 5, http://treasury.gov.au/carbon-pricemodelling.

31 "100 TWH of clean-energy facilities ...": Note that 100 TWH of "energy" is generated over the course of a year from the sustained supply of 11,400 megawatts of "power."

31 "we generate about 9 TWH ...": We also generate a significant amount of hydro-electric power but this is not a candidate for rapid expansion. See Australian

Bureau of Agricultural and Resource Economics and Sciences, *Energy in Australia 2011*, ABARES, Canberra, 2011.

32 "it will deliver around 0.4 TWH of energy ...": This assumes that solar power is operating at 30 per cent capacity, which is consistent with international experience. See http://iga.igg.cnr.it/documenti/IGA/Fridleifsson_et_al_IPCC_Geothermal_paper_2008.pdf.

32 "building five wind turbines every single day ...": Based on 19,000 2-megawatt turbines operating at 30 per cent capacity, delivering 11,400 megawatts of power which is equivalent to 100 TWH of energy over a year. Thirty per cent capacity is a reasonably optimistic operating assumption for a standard wind turbine.

32 "twenty standard 750-megawatt reactors ...": Views on the nuclear requirement vary. Professor Leslie Kemeny wrote in the *Australian Financial Review* on 17 September 2011 that five 1-gigawatt nuclear power stations would be sufficient to reach a 5 per cent reduction in Australian emissions.

33 "Nuclear power plants are expensive ...": Assuming reactors operate at 80 per cent capacity all year round.

33 "The Treasury documentation states ...": Australian Treasury, 21 September 2011, http://treasury.gov.au/carbonpricemodelling/content/report/09chapter5.asp.

33 "two-thirds of the 'carbon pollution reduction' achieved ...": In a last-minute concession to the Greens, the final version of *Clean Energy Future* policy documentation indicates a preference to limit foreign carbon credits to half of all permits. The Greens trumpeted this as a success, claiming that the agreement "limits the use of international offsets (both quantity and quality)." This is nonsense. The new requirement merely requires that no more than 50 per cent of all permits be purchased from abroad. This does not mean that 50 per cent of reductions come from abroad. In fact, that requirement is so weak that the Treasury modelling indicates it will have no effect: "in all policy scenarios ... the cap does not bind." See Australian Treasury, 21 September 2011.

34 "Nearly 3 billion people still cook on indoor stoves ...": IEA, *Energy Poverty: How to Make Modern Energy Access Universal*, OECD/IEA, Paris, 2010.

34 "Smoke from indoor stoves and kerosene lighting ...": World Health Organization, *The Global Burden of Disease*, WHO, Geneva, 2008.

36 "Griffith has done the maths ...": Griffith estimates the world will need to generate 17.5 terawatts of power (150,000 TWH of energy per year). To soften the impact of climate change (keeping atmospheric concentration to 450 parts per million) in the next twenty-five years, Griffith says we will need to reduce our fossil-fuel use to 3 terawatts, which means generating 13 terawatts of clean power in addition to the 1.5 terawatts of biofuels and nuclear that we already

# Never again miss an issue. Subscribe and save.

☐ **1 year subscription** (4 issues) only $49 (incl. GST). Subscriptions outside Australia $79. All prices include postage and handling.

☐ **2 year subscription** (8 issues) only $95 (incl. GST). Subscriptions outside Australia $155. All prices include postage and handling.

☐ Tick here to commence subscription with the current issue.

**PAYMENT DETAILS** Enclose a cheque/money order made out to Schwartz Media Pty Ltd. Or please debit my credit card (MasterCard, Visa or Amex accepted).

CARD NO. ☐☐☐☐ ☐☐☐☐ ☐☐☐☐ ☐☐☐☐

EXPIRY DATE        /                    AMOUNT $

CARDHOLDER'S NAME

SIGNATURE

NAME

ADDRESS

EMAIL                                    PHONE

**tel:** (03) 9486 0288  **fax:** (03) 9486 0244  **email:** subscribe@blackincbooks.com  **www.quarterlyessay.com**

---

# An inspired gift. Subscribe a friend.

☐ **1 year subscription** (4 issues) only $49 (incl. GST). Subscriptions outside Australia $79. All prices include postage and handling.

☐ **2 year subscription** (8 issues) only $95 (incl. GST). Subscriptions outside Australia $155. All prices include postage and handling.

☐ Tick here to commence subscription with the current issue.

**PAYMENT DETAILS** Enclose a cheque/money order made out to Schwartz Media Pty Ltd. Or please debit my credit card (MasterCard, Visa or Amex accepted).

CARD NO. ☐☐☐☐ ☐☐☐☐ ☐☐☐☐ ☐☐☐☐

EXPIRY DATE        /                    AMOUNT $

CARDHOLDER'S NAME                       SIGNATURE

ADDRESS

EMAIL                                    PHONE

RECIPIENT'S NAME

RECIPIENT'S ADDRESS

**tel:** (03) 9486 0288  **fax:** (03) 9486 0244  **email:** subscribe@blackincbooks.com  **www.quarterlyessay.com**

**Delivery Address:**
37 LANGRIDGE St
COLLINGWOOD  VIC  3066

Quarterly Essay
Reply Paid 79448
COLLINGWOOD  VIC  3066

---

**Delivery Address:**
37 LANGRIDGE St
COLLINGWOOD  VIC  3066

No stamp required
if posted in Australia

Quarterly Essay
Reply Paid 79448
COLLINGWOOD  VIC  3066

have. Griffith is quoted in Stewart Brand, *Whole Earth Discipline*, Viking Press, New York, 2010. The calculations above are presented at http://web.me.com/stewartbrand/DISCIPLINE_footnotes/Griffith_slides.html. Brand indicates that the calculations are based on a combination of Griffith's published speaking materials and his own conversations with Griffith.

37    "installing 100 square metres of solar cells every second ...": Assuming 15 per cent efficiency.

37    "requiring 50 square metres of reflective mirrors ...": Assuming 30 per cent efficiency.

40    "fastest-growing global energy source ...": IEA, *Clean Energy Progress Report*, OECD/ IEA, Paris, April 2011.

40    "Chinese coal consumption is increasing ..." Robert Kaufmann et al., "Reconciling anthropogenic climate change with observed temperature 1998–2008," Proceedings of the National Academy of Sciences USA, 2011.

40    "without clean coal, the overall cost of tackling climate change ...": IEA, *Carbon Capture and Storage Model Regulatory Framework*, OECD/IEA, Paris, 2010.

42    "Wind energy is around 50 per cent more expensive ..." These figures reflect "levelised costs" – a comparable summary measure of the overall competiveness of different generating technologies which takes into account the total cost of building and operating a generating plant over an assumed financial life. See "Levelized Cost of New Generation Resources in the Annual Energy Outlook 2011," US Energy Information Administration, December 2010, http:// 205.254.135.24/oiaf/aeo/electricity_generation.html.

44    "Americans placed global warming last ...": Pew Research Center, *Public's Priorities for 2010: Economy, Jobs, Terrorism*, 25 January 2010.

45    "to 2 degrees Celsius compared with pre-industrial levels ...": If we allow temperatures to rise by 5 degrees or more – higher than at any time in the last 30 million years – we would suffer extraordinary impacts on our agriculture, coastlines and weather.

45    "greenhouse gases in the atmosphere must be kept below 450–500 parts per million ...": If we continue on our current path, greenhouse gases will rise to around 750 ppm or three times pre-industrial levels by the end of the century.

45    "reduce global emissions by at least 50 per cent ...": That is 50 per cent below 1990 levels. To do this we'll need to reduce emissions in rich countries by more than 50 per cent to allow for a smaller cut in developing countries.

45    "successful application of carbon-capture technology to coal plants ...": Whether CCS is feasible and, even if it is, whether retrofitting is practical, is a highly controversial issue.

45 "between two and four times slower than what is required ...": IEA, *World Energy Outlook 2010: Fact Sheet*: "carbon intensity – the amount of $CO_2$ emitted per dollar of GDP – would have to fall at twice the rate of 1990–2008 in 2008–2020 and four times faster in 2020–2035."

46 "Kyoto Protocol, despite all the fanfare ...": Gwyn Prins & Steve Rayner, "Time to ditch Kyoto," *Nature*, Vol. 449, 25 October 2007, pp. 973–5.

46 "offshoring them to manufacturing plants ...": Andrew Brinkley & Simon Less, "Carbon omissions: consumption-based accounting for international carbon emissions," *Policy Exchange*, 19 November 2010.

50 "every year between 2010 and 2050 the world will need ...": IEA, *Energy Technology Perspectives*, OECD/IEA, Paris, 2008, p. 42.

50 "$100–500 per tonne, not the $20–30 now planned ...": IEA, *Energy Technology Perspectives 2008: Fact Sheet – The Blue Scenario*. Achieving a 50 per cent reduction in emissions requires "a cost up to USD 200/t. If technology progress is less than we expect, the cost may rise to USD 500/t."

51 "solar projects have achieved falling costs ...": Ross Garnaut, *Garnaut Climate Change Review*, "Update paper 7: low emissions technology and the innovation challenge," Australian Government, 2011.

54 "trialling a limited emissions trading scheme in some provinces ...": China's latest Five-Year Plan features targets to cut energy intensity (energy consumed per unit of GDP) by 16 per cent and carbon intensity ($CO_2$ emitted per unit of GDP) by 17 per cent by the end of 2015.

61 "humans are an intrinsic element of a living earth ..." Tim Flannery, interview, *The Science Show*, ABC radio, 1 January 2011. Flannery believes that "within this century the concept of the strong Gaia will actually become physically manifest" when "This planet, this Gaia, will have acquired a brain and a nervous system. That will make it act as a living animal."

61 "Dick Smith warns ..." Dick Smith, *Dick Smith's Population Crisis*, Allen & Unwin, Sydney, 2011, p. 197. The full quote is: "The endless growth economy is obsolete and risky to future generations. We must plan now and begin to implement a steady state economy based on quality of life rather than quantity of consumption." On p. 190 Smith claims that the current rate of economic growth "is likely to overwhelm any foreseeable advances in technology or potential efficiencies."

# THE AUSTRALIAN DISEASE

2011 *Alan Missen Oration on the Decline of Love and the Rise of Non-freedom*

Richard Flanagan

Good evening. I am afraid Richard Flanagan couldn't be here tonight and so I am here in his place. My name is Craig Thomson. You might have heard a lot in recent times about another Craig Thomson. I am not him. I am just a run-of-the-mill bloke who likes prostitutes. A lot. They're great. Really, really great. God I've had some fun. You can do anything if you pay. It's incredible. I even had a Julie Bishop once. She said, You want the stare? I said, You get what you pay for.

Mates say, But how do you afford it? Mate.

I say, you find someone who has your name and a work credit card with a huge limit and you work out where this person is staying and you break into his room. Or just knock. That's what I ended up doing. We got to know each other quite well, and well, he's so busy doing his job, I just borrow his wallet and his phone and ring up and do the ordering.

This is an edited version of the 2011 Alan Missen Oration, run by Liberty Victoria and doubling as the closing address of the 2011 Melbourne Writers' Festival. It draws on essays in my most recent book, *And What Do You Do, Mr Gable?* to which interested readers are referred.

He never seems to mind, he doesn't really seem to know who I am – it's sort of like a Philip K. Dick novel, doppelgangers and whatever – and then you get down to business. Christ, I was there once, video camera and all, but he didn't seem to notice. He'd be talking to someone important. Like maybe Mark Arbib. Or somebody like that. Perhaps it was another universe or parallel dimension, the ALP, I don't know. All I know is that when I see Craig now he doesn't seem to remember me.

Anyway, Richard Flanagan asked me to read his speech and I'm hoping to borrow his credit card later tonight, so I should begin.

*

How good it is to be here at the end of a marvellous fortnight that has seen literature celebrated in the most wonderful way possible. Whose heart hasn't lifted with joy at all that has taken place in our great city of fiction?

I speak, of course, of Canberra.

In the offices and lobbies and forecourts of Parliament House, the restaurants and bars of Manuka and Kingston, homage has been paid by our leaders to literature in the most sincere way possible, by life seeking to imitate art.

For on this, the fiftieth anniversary of *Catch-22*'s publication, how inspirational it is to see a character clearly modelled on Joseph Heller's Major Major Major Major, forever absent and suddenly somehow supremely present, promoted by bureaucratic error that is the fault of a machine gone haywire – in Major Major Major Major's case an early IBM computer, in Craig Thomson's case a late NSW Labor machine – to centre stage of our nation's public life.

And that, of course, is the catch. Because Craig Thomson shouldn't have been chosen and promoted by the ALP – except that he was and now, of course, he can't be unchosen.

The question arises as to why the ALP endorsed someone like Craig Thomson for the last election, when much of his past would presumably already have been known in ALP circles. And having elevated him, why did the ALP keep promoting him? Why did Craig Thomson get the thumbs up from the Labor Party, why did the Labor Party think this man should be a leader, a lawmaker, a shaper of opinion and of our national future?

And the answer, perhaps, is one that goes to the heart of our present malaise: it is because Craig Thomson was a conformist. And in present-day Australia, it doesn't matter what you do or what you have done, so long as you conform to power. The only true crime in an ever-more bland Australia is to not conform. I don't mean to suggest that conformity is a national characteristic. It is an aspect of the human condition, but only one part. But of late it seems to have become a predominant condition of Australian life. And it is about this Australian disease of conformity that I want to speak tonight, and about how conformity deforms and destroys love and freedom.

Let me begin with a story my father once told me. During the war he was one of "Dunlop's Thousand," that now mythical group of POWs who endured the horrors of the Death Railway under the Japanese, led by a doctor called Weary Dunlop.

One day on the railway a digger called Slappy Oldham turned up to sick parade with a cigarette dangling from the corner of his mouth. An English major called Driscoll made a swipe at Slappy, which the POW evaded by the slightest move of his head.

"Lucky you missed," said Slappy Oldham.

Driscoll angrily demanded to know why.

"If you'd touched me," said Slappy Oldham, "I'd have dropped you, you bastard."

Driscoll grew more agitated and was speaking of charges when Dunlop arrived. Slappy walked up to him.

"You know that bastard Driscoll?" said Slappy Oldham. "He tried to swipe me, and I told him off."

"Good on you, Slappy," replied Weary, to the amazement of the upper echelon. "Always look after yourself."

I once spent a memorable evening with Dunlop that ended with us drinking in his Toorak mansion, a place where time seemed to have stopped somewhere in the 1930s. I sat on an aged, cracked leather couch and he told me of how as a young doctor in London in the 1930s he had gone into the East End and taken on Oswald Mosley's British Union of Fascists' Blackshirts at their rallies. I knew this was unusual, for Dunlop was a distinguished rugby player, capped for Australia, and Mosley recruited heavily from London rugby circles. So concerned was Dunlop by the rise of fascism, he told me, that he very nearly went to Spain to fight with the International Brigades.

Was he attracted to the Communist Party then? I asked, knowing well how many of the very best at the time had been.

"Not at all," Dunlop said. "I just didn't agree."

Dunlop spent his war not agreeing, his methods often involving literature. He said the best way to deal with a Japanese guard was by learning Shakespeare's sonnets. His own weapon of choice when seeking to gain the upper hand in a cat-and-mouse game of asserted authority was Portia's speech from *The Merchant of Venice* on the quality of mercy, which he would declaim in a loud and forceful enough manner to bluff his way out of trouble. Sentenced to death by beheading, he passed what he thought were his final hours reciting Keats' "Ode to a Nightingale." The central idea of Keats' most famous poem is that the singer may die, but the song is eternal. Dunlop's choices were ever apt.

In his old age the poem that meant most to him was Tennyson's "Ulysses," a poem of an aged king reflecting on what has come to him.

"I am become a name," thinks Ulysses. And so too, Weary Dunlop.

A letter he wrote my father ends with the final stanza of "Ulysses," in which the aged king implores his old colleagues – "free hearts," as he calls them – to follow him back to sea, to

> ... sail beyond the sunset, and the baths
> Of all the western stars until I die. ...

Tho' much is taken, much abides; and tho'
We are not now that strength which in old days
Moved earth and heaven; that which we are, we are;
One equal temper of heroic hearts,
Made weak by time and fate, but strong in will
To strive, to seek, to find, and not to yield.

The poem, particularly this final stanza, is an idea of freedom, of resistance at once utterly futile and somehow entirely necessary, and that idea is in literature closely related to the idea of love. *Catch-22*, a novel of heroic resistance in which the hero recognises he is jeopardising his traditional rights of freedom and independence by daring to exercise them by taking part in a war, begins, like so many novels, with love, Heller's opening sentence reading, "It was love at first sight." In the next paragraph we discover Yossarian has fallen in love with the chaplain.

Love, humour, freedom. From the first great recognisably modern novel, *Don Quixote*, these three are never very far apart. Do we trust humour because it is the justice the law never is? Do we trust the idea of love, no matter how mad – from Don Quixote's love for the farm girl he renames Dulcinea del Toboso to Yossarian's love for the chaplain – because in the face of conformity, the madder the love, in some mysterious way the greater the commitment to freedom?

"Freedom, Sancho," Don Quixote says, "is one of the most precious gifts bestowed by heaven on man; no treasures that the earth contains and the seas conceal can compare to it; for freedom, as for honour, men can and should risk their lives and, in contrast, captivity is the worst evil that can befall them."

Freedom and love. Both are complex and irreducible and ambiguous. Or, to put it more simply, both are deeply human. And correspondingly, their denial is an attack on what we most value about ourselves. Yet to discover their essence, to live these things, is the mystery with which literature is so often obsessed.

"For one human being to love another," wrote the poet Rainer Maria Rilke, "that is perhaps the most difficult of all our tasks, the ultimate, the last test and proof, the work for which all other work is but preparation."

Few writers of the last hundred years have more bravely attempted to describe the growing difficulty of that task of love than the Soviet writer Vasily Grossman in his last two great novels, Life and Fate and Everything Flows, in which Grossman seemed to conclude that politics was the enemy of love.

By 1962, Grossman was dying of stomach cancer. Once one of the most celebrated Russian war journalists and an acclaimed social-realist novelist, Grossman was now in disgrace. The KGB had confiscated all known manuscripts of Grossman's new epic novel of World War II, Life and Fate. Grossman met with Mikhail Suslov, chief ideologue of the USSR's Politburo, to beg that his book be published.

Grossman was told that his novel was more damaging to the USSR than Pasternak's recent cause célèbre, Dr Zhivago. It could not be published, said Suslov, for 200 years.

Unknown to either Grossman or the KGB, one of Grossman's friends had made a secret copy. Nearly twenty years after Grossman's death, it was smuggled out of the USSR. Though it made little impact on publication in Switzerland in 1980, it has in the decades since come to be hailed as a twentieth-century War and Peace, and with this changing fortune Grossman has secured a reputation as a latter-day Tolstoy.

All praise is a form of incomprehension, and Grossman is a writer more difficult than most to divine. Unlike the great Soviet writers who were products of pre-revolutionary Russia – like Bulgakov and Akhmatova – Grossman was a product of the new order, an insider, a Soviet man.

"Everything I have," he wrote in a letter to the NKVD chief Yezhov in 1938 begging for the release – really, the life – of his arrested wife, "I owe to the Soviet government."

Why this man – a conformist who made his accommodations with the Soviet tyranny, turned his back, averted his eye, held his tongue, signed accusing letters – came to a point where he said "No" to his masters is perhaps unknowable.

Certainly his experience of the war, his witnessing of the Holocaust, the death of his mother at the hands of the Einsatzgruppen, Stalin's post-war anti-Semitic campaign, his discovery of love in middle age – a large life, in short, that cannot be detailed here – led Grossman finally to conclude that fascism was simply a mirror response to the "cosmic violence" of Soviet communism. But why this then liberated him to write two master-pieces of the twentieth century remains mysterious.

A few months after his meeting with Suslov, at the height of the Cuban missile crisis on 26 October 1962, the Central Committee heard that Gross-man was at work on a new "anti-Soviet" novel. The informer is suspected to have been his stepson, who lived with Grossman.

As the world teetered on the brink of the abyss, this novel was discussed at the highest levels of Soviet leadership. Is it possible to imagine any book today eliciting such fear, bequeathed such strange honour?

The novel was Everything Flows, unfinished at the time of Grossman's death in 1964. It was perhaps unfinishable. How strange it is though that somewhere in that abyss between ambition and failure sometimes lies greatness. When first published in English in 1972, Everything Flows failed to garner anything like the attention in the West that Solzhenitsyn's work had received. Grossman's idea of history was heretical to almost all. He didn't compare or rank the horrors of the gulag and collectivisation and the Holocaust. Rather, and most chillingly, he connected them.

His anti-politics, of a type that anticipated the great revolts of the 1980s, rendered his work unfathomable for many. The book offered succour neither to the left, for by reviling Lenin and Leninism it broke the ultimate taboo; nor to the right, by offering a damning critique of pre-revolutionary Russia.

His humanism, placing kindness and goodness, truth and freedom, at the centre of life, as both the meaning and fullest expression of life, seemed weak, even quaint, in the face of the cocaine rush of turbo-capitalism that had begun to take all before it in the final decades of the twentieth century with its material wonders and ideological triumph.

In recent years, particularly since the global financial crisis, Grossman's *Everything Flows* has become the subject of worldwide interest and praise. What has changed is not Grossman, but us. Suddenly, this story seems not about another world many years ago, but speaks to our world now and tomorrow.

The novel is ostensibly simple and could hardly be simpler. In 1957, after thirty years in the gulag, a man returns to a humble life in Russian society. Within it, though, is a book constantly breaking boundaries, flooding over, travelling far from the strange anti-socialist realist, social-realist *Life and Fate*; pointing to the great philosophical novels of Kundera; constantly keeping faith with the idea of story as the vehicle of truth.

The book contains multitudes, and not only of people. Its moods range from the near mystical, in its depiction of women, particularly mothers, to hard political, in its study of Lenin, to epic and elegiac. Grossman somehow penetrates the essence of the USSR in a way few ever did – alive to the psychology and the humanity of its revolutionaries, cannibals, *zeks*, commissars and secret policemen. His stark descriptions of the horrors of the Ukraine famine make one shudder today; I suspect they will have the same effect in centuries to come.

"All the living are guilty," he writes, a judgment he did not exclude himself from. He had signed the letters, he had refused to help, and he bore especial guilt about his mother, whom he felt he should have saved.

The dying Grossman is a novelist now going for broke. Like the dying Bulgakov writing *The Master and Margarita*, he was liberated from fame, success, even the possibility of publication, to be able finally to write what he meant.

Near its end *Everything Flows* breaks its banks again and again: chapters grow shorter, more concentrated, reducing history, thought, human nature, to a dazzling and dizzying poetry.

Grossman makes a chilling historical argument replete with the ultimate Soviet blasphemy, the essence of which is still shocking to come to terms with. That it was written half a century ago makes it even more extraordinary. Grossman argues that the nineteenth-century prophets of the unique Russian soul, from Gogol to Dostoevsky, believed that this soul, once fully realised, would lead the world to a profound spiritual evolution. Their fatal error, according to Grossman, was that they "all failed to see that this soul had been enslaved for a thousand years."

For Grossman, Russian history was that of a thousand years of slavery. He traces the growing enslavement of the Russian people through the Middle Ages and argues that the progressive achievements of Peter the Great and Catherine the Great were linked to a corresponding increase in what he calls "non-freedom."

In Russia Grossman sees progress and the growth of slavery as inextricably linked, while in the West it is progress and the growth of freedom. And here is the heart of Grossman's terrifying vision: the true consequence of Lenin's revolution was to take this uniquely Russian slavery to the world. This for Grossman is the Russian spectacle that enchants the world: "of modernisation through non-freedom."

For Grossman, Stalin is but a consequence, and an inevitable one at that, of Lenin. And not just Stalin, but fascism. Writing this in the early 1960s in Russia was more than merely blasphemous. It was an historical insight of extraordinary perceptiveness.

"Did Russia's prophets ever imagine," Grossman wrote in the final months of his life, "that their prophecies about the coming universal triumph of the Russian soul would find their grating fulfilment in the unity of the barbed wire stretched around Auschwitz and the labour camps of Siberia?"

Grossman's hero, Ivan Grigoryevich, senses the spirit of the gulag all around him. "Barbed wire, it seemed, was no longer necessary; life outside the barbed wire had become in its essence no different from that of the barracks."

This terrified me when I first read it; that an idea had escaped the gulag and might take the world. I am still shaken by it.

The great irony, according to Grossman, was that Lenin, through his violence and terror, destroyed not just any possibility of liberation from "the satanic force of Russia's serf past," but hugely advanced its domain. And so, writes Grossman, "Through the will, passion and genius of Lenin, Russia's thousand-year law of development became a worldwide law."

And who looking at China can read this without trembling? Who can contemplate the USA's present stumblings and the rise of the Tea Party movement without wondering? Was the outcome of the Cold War defeat and victory, or was it a strange merging of the worst of two polarities to form our brave new world? The left believes freedom will bring economic progress, the right that economic progress will bring freedom. But what if both are wrong, if Grossman is right, and there is no necessary connection?

"A Putin–Palin ticket," recently suggested Gary Shteyngart – another Russian-Jewish writer whose satires focus on turbo-capitalism's similarities to old-style totalitarianism – "can really cement the liberties Russia has achieved over the last 200 years."

Yet as he journeys through hell, somehow Grossman divines meaning in all this, and is finally hopeful. He concludes that freedom can never be destroyed. For Grossman this is "a sacred law of life: human freedom stands above everything. There is no end in this world for the sake of which it is permissible to sacrifice human freedom."

Half a century after his death it is Grossman's twin insight – his assertion of freedom over politics coupled to his fear of an idea escaping the gulag and taking over the world, this terrifying vision of material progress achieved through non-freedom – that resonates. For we assume freedom when perhaps we are not entirely correct to do so.

Non-freedom to the Western mind is inevitably linked with images of backwardness – Soviet tractors, East German Trabants, Kim Jong Il's haircut. But non-freedom these days is also iPads, iPhones and a dazzling array of less iconic but ubiquitous consumer goods that flood our stores, our homes and which increasingly are used to define our ideas of worth and happiness. It is a full-lipped smile achieved with the aid of collagen made from skin flensed from dead Chinese convicts.

And over the last two decades the idea that escaped the gulag has begun appearing in other more direct ways in Australia. And strangely, there is a growing tolerance of and even active support for these manifestations of non-freedom.

Take, for example, the anti-terrorism laws passed several years ago. Supported by both parties but much and rightly condemned for their draconian and utterly unnecessary provisions for secret trials and secret imprisonment, these laws are profoundly dangerous. As long as they remain on the books an unscrupulous government in difficult circumstances could visit great evil on Australian people. They need reform as a matter of the greatest urgency and yet there is no likelihood they will be reformed.

In the Dr Haneef case we were presented with the disturbing spectre of an innocent man appearing to be framed and imprisoned for what can only be seen as political advantage, in which our security forces would seem to have lied to the Australian public. In 2007 a Sydney court found that an Australian citizen, Izhar Ul-Haque, had been illegally kidnapped and threatened by ASIO operatives. This blatant abuse of power by our secret police and what it might bode for our future seemed of little concern to either major party or the media, and the lack of attention the case received was all the more remarkable given it was held in the middle of the federal election campaign.

A second example is the shameful decision by the federal government, days after it recorded its worst polling on record, to seek to seize David Hicks' earnings from his memoirs. While Chopper Read can boast of feeding people into cement mixers on national television and derive income from his highly successful books, it seems strange in the extreme that a figure whose criminality is a matter of national debate rather than established fact would be pursued by the government in this matter.

It is very hard to escape the suspicion that, having lost a large section of the voting public with its carbon tax, the government sought a few weeks later to win it back with a right-wing play. Why otherwise would the government wait for nine months after a publication that the federal police tracked every step of the way?

David Hicks' supporters contend that Hicks was effectively a political prisoner at Guantanamo. If they are right, then rather than the government seeking to perpetuate the impression Hicks is a criminal by taking money he has earned from his memoirs, the government itself deserves public investigation as to the extent to which it was complicit in allowing Hicks to be tortured and imprisoned because of politics. And for any such investigation to have validity, Hicks' own testimony needs to be widely published and discussed.

If Hicks' supporters are wrong, this deserves to be established, with both sides of the argument being made thoroughly by all involved, not least and most particularly by Hicks himself. Hicks's story was a central one for Australia in recent years and the issues raised are profound for any country that perceives itself as democratic. They therefore deserve the widest and deepest consideration. To seek to seize the profits from his book does not promote that consideration but rather seeks to stifle it.

As the conservative commentator Chris Berg noted, the argument that confiscating David Hicks' commercial gains has nothing to do with free speech is misleading.

"That this argument has come from many conservatives is disappointing," writes Berg, who goes on to argue: "Speech requires finance. To pretend the former is unharmed if you ban the latter is nonsensical. If a government was to ban a newspaper from making a profit but otherwise leave its material uncensored we would not hesitate to condemn it as a violation of freedom of expression."[1]

While the federal government's decision then seems explicable only in terms of the pathetic calculus that is contemporary Australian politics, it creates worrying precedents about the possibilities of future political censorship through the impoundment of income. The inescapable lesson of history is that any power given to the state – even when well meaning in intent – will ultimately be abused by the state. Like freedom, non-freedom does not arrive ready-made. It grows, and no soil should ever be made ready for its sowing.

A third area of growing non-freedom is our treatment of boat people. If fifty Australians drowned in Sydney Harbour, it would be a national tragedy. But when upwards of fifty refugees drown off the Australian coast, as they did at Christmas Island late last year, it is a political question. Even worse than the varied Pacific laagers proposed by successive governments is the laager of the mind in which both major parties are now incarcerated. Sadly, a Nauru of mentality is immune to judicial review. The racial and social panic that was whipped up to win an election in 2001 has now taken hold to such an extent that neither party can seem to conceive how to approach the matter without resorting to more hysteria.

If Australia does not have a refugee problem like, say, Italy, with over 48,000 refugees arriving since January, it does have a dismal public life largely bereft of courage or humanity, and it has created a national myth that now poisons all sides of politics. The myth is that hordes of refugees will overrun Australia unless harsh policies of dissuasion and internment are employed.

For more than a decade this myth, the issue of opportunism and electoral cynicism, has been a weeping sore at the heart of our public life. It cannot be blamed solely on John Howard's Liberal government. It was the product of a deep cynicism in both major parties and has since become deeply entrenched in no small part because – despite some honourable exceptions – of the lack of courage within either major party to stand up against it.

The strange drift begun in the 1990s under Paul Keating with detention centres and mandatory detention centres was immensely strengthened under John Howard's government. As loony as Hitler's *Lebensraum* in reverse, John Howard's rhetoric of "border protection" struck a similarly popular note. The Labor Party, as noted by US embassy officials in a

WikiLeaks cable published the same day as last year's Christmas Island tragedy, was profoundly traumatised by its 2001 election loss and remains haunted by it to the present day. Labor's leadership capitulated to Howard's vision and went largely in lock step with his policies, and on coming to power Julia Gillard has essentially reworked Howard-era policies with a sometimes ludicrous edge.

Wicked as it was, the myth grew only more powerful as the rate of suicide, of self-harm, of simple and utterly unnecessary waste and wreckage of human life among the refugees ballooned. And perhaps even more damaging was the harm we did ourselves. Numerous psychological studies have demonstrated how human beings can be desensitised to the sufferings of others, how empathy can be eroded to the point where otherwise reasonable people can inflict great suffering in good conscience. Australia over the last two decades has been one vast psychological study in which our leaders have desensitised a nation to the plight of others. There is no good in this, and the portents of where it is leading us in the form of the rise of racist attacks and hate websites, and the growing influence of a new far right, are disturbing.

What does it say of a nation, what does it say to a nation, when in a time of austerity, of slashing of public services, a billion dollars of Australian taxpayers' money is being spent annually to persecute, damage and sometimes destroy the lives of people of whom between 80 and 95 per cent are finally proven to be genuine refugees – that is, to hurt the most powerless and helpless and deserving of help and kindness? It shames us as a nation that claims to be both humane and generous, it belittles us as a people, and none of it will deter the wretched of the earth, forced to choose between despair and hope, from continuing to choose hope.

On 1 September 2011, the prime minister, Julia Gillard, looking and sounding like a dying metronome, criticised the High Court decision that halted the incompetently conceived and ineptly named Malaysian Solution. Admitting that there was now doubt over whether offshore processing was any longer legally possible, she went on to say of refugee law that

"The High Court's decision basically turns on its head the understanding of the law in this country."

It would, though, be too much to hope she's right, to hope it is the beginning of the end of this poisonous and cruel fiction that has for far too long caused so much misery for so little reason, because the one bizarre certainty is that both parties will only harden their rhetoric and redouble their attempts to find a way to ignore or overturn the High Court's decision.

For much of the latter part of the twentieth century Australia seemed to be opening up to something large and good. It believed itself a generous country, the land where battlers got a "fair go." Whatever happened to a fair go? Whatever happened to the battler? Because if an Afghan Hazara isn't a battler, I don't know who is. In the video footage of last year's tragedy it is possible to look down the cliffs of Christmas Island and in the spindrift blown up from below to hear not only the screams of women and children, to see not only the drowned and the drowning and a broken boat, but also to glimpse the promise of what Australia had once been. And with each wave that rolls in, it breaks apart a little more.

And watching this, it is too easy to deride Howard, to dismiss Gillard, to mock Abbott. Far harder to understand the larger drift of our times that has led our politicians to behave so abysmally, to recognise that neither our problems nor their solutions come with party tags. For we have reached a moment in history where politics seems suddenly unequal to the terrible problems that beset us.

If we look at Australia over the last decade and a half, we are presented with the unedifying, indeed disturbing, image of a society whose major institutions failed. This was not necessarily so in other countries. If it is the case, for example, that the past US administration committed crimes – at Abu Ghraib, at Guantanamo, in rendition centres – then it was US journalists who first brought them to public light, it was US legal systems and US lawyers that began bringing them into question, it was US public figures who began pressing for change. Nothing similar happened in Australia.

If we look to another example, that of Britain, we see that it had what we didn't: a major debate in parliament about whether they should go to war in Iraq. On the tombstone of the former British foreign affairs minister, Robin Cook, who resigned his parliamentary positions over the Iraq invasion, are his own poignant words: "I may not have succeeded in halting the war, but I did secure the right of parliament to decide on war." One after another, loyal Tory MPs and loyal Labour MPs stood up and said they disagreed with their party leadership's support of the war.

To the shame of all Australian parliamentarians, not one here could claim the same epitaph as Cook, for in contrast our parliament was quiescent. In Australia such a questioning of the party leadership's position on any issue has become not just unacceptable, but pilloried in the impoverished judgment of the Canberra press gallery as political suicide. To speak out is to be declared a rat, a party renegade and a political naif. To not speak out is to be rewarded with endorsement and promotion. It is to be Craig Thomson. It is the Australian disease.

And so our parties failed us. Our parliament failed us. Our media failed us. Why they did is difficult to answer, though the answer is clearly to be found in an examination of the new conformity at the heart of Australian life.

What we have witnessed is a corrosion of the idea of the truth and a sustained attack on those whose views differ from that of power. What we have experienced is a coarsening of public rhetoric by standover men who claim to speak for the ordinary Australian, but seem rather to represent only the interests of government and corporate power. They are the belligerati who are given opinion columns and radio talkback programs; who are accorded the status of minor celebrities; and there can at times seem no end to the uniquely Australian cross of their public bullying in defence of private interest at popular expense.

When the Velvet Revolution took place, Václav Havel said the West was wrong to dismiss the experience of Eastern Europe as history. Rather, he said, it was a distorted mirror to what the West could become if it wasn't vigilant.

What we hear parroted in Australia today at ever-shriller frequencies are the old mantras of Stalinism, once used to justify the great crimes of a century, now ironically recycled by the right to defend the indefensible. Those who speak out are inevitably demonised as out-of-touch elites. This pejorative use of the word elite begins with Stalin in 1948, when he used it to describe Jewish intellectuals upon whom he was about to turn his terror.

We are being told, as the old USSR was told, that there are things that matter more than the truth and individual freedom – national security, the needs of the security forces, special international commercial undertakings. But as Vasily Grossman came to conclude, there is nothing higher in this life than the truth and individual freedom. The striving for these two things is the essence of who and what we are.

One of the most potent expressions of our new age is the arrival of a new class, both political player and truth controller, composed of those individuals for whom the role of politician, journalist, minder and senior bureaucrat are just avatars they inhabit, interchangeable ways of exercising power against truth and against freedom.

The phone-hacking scandal, with its revelations of revolving Rubik's cubes of power involving media, politics and police, was an illuminating insight into this new class, which resembles nothing so much as the old class of apparatchik in the Soviet Union, mediocre vessels empty save for vaulting ambition and endlessly craven souls. It's not just that Rupert Murdoch seems to be morphing into Yuri Andropov in a leisure suit. Conformists par excellence, capable only of agreeing with power however or wherever it manifests itself, they are the ones least capable of dealing with the many new challenges we face, precisely because those challenges demand the very qualities the new class lacks: courage, independence of thought and a belief in something larger than its own future.

The new class, understanding only self-interest, believing only in the possibilities of its own cynicism, committed to nothing more than its own perpetuation, seeks to ride the tiger by agreeing with all the tiger's desires, believing it and not the tiger will endure, until the tiger decides it's time to feed, as the mining corporations did with Kevin Rudd, as News Limited is now with Julia Gillard.

Nor does the new class have any answer other than accommodation to the rise of a new far right unseen in the West since the 1930s, a strange and inchoate rage waiting for final and terrible political expression. At a moment in history when the old verities are crumbling, this new far right trades in anger and fear, in conspiracies and known enemies. It promotes a new consensus that celebrates superstition as knowledge, that is hostile to social and racial difference, and celebrates hatred at the expense of reason. It leads to scientists being not just derided, but routinely threatened with death, and intellectual and artistic life seen as at best unnecessary and at worst dangerous. And all because this new rabid right is interested not in truth, but promoting ignorance; not in freedom of all, but the righteous punishment of those it regards as the damned. Rather than recognising the complex task of building lives together in a world of difference, its rhetoric promises an apocalyptic liberation.

Under the rule of the new class, we remain smug and complacent. We ignore the worst global economic crisis since the 1930s as the folly of far lands. We understand a once-in-a-century resource rush as an immortal fact of our own country. We confuse robbing the wealth of our land with an idea of national genius. We mistake corporate success for personal prosperity. Yet a few days after BHP announced a record profit of $22.5 billion, Australia's biggest ever, ABS statistics were released that show Australians' total disposable income fell for the first time in fourteen years. Our gilded recession may well at a certain point become a full-blown recession. Or worse.

Are we ready for what we must deal with? Do we have the political will, the collective empathy, to deal with the social stresses that will inevitably arise without resort to more attacks on our liberties?

We in Australia should not make the error of thinking that the causes of the horrors besetting other countries are unique to them. The hatred and fear of Muslims that motivated Anders Breivik to massacre seventy-six Norwegians is a hatred and fear that is prospering in Australia. The attempted assassination of a US senator earlier this year is an idea of change being given legitimacy by the careless language of both politicians and some sections of the media here in recent months.

We need to remind ourselves that material progress, corporate profits and a mining boom do not need freedom to happen. We should not forget that when BHP was heading towards the seventh-biggest profit in the world's history it did not hesitate to destabilise our elected prime minister for proposing to use a fraction of that profit for the national good.

Democracy suffers most when it is wrongly presumed that its guarantees are to be found in the state or government or party, in history or myths of national goodness. Democracy may be the best antithesis to tyranny, but it is not necessarily wise or good. It can at times be an obscene spectacle guilty of great and historic crimes, that on occasion slaughters its own and others for no good reason, colluding with corporations and corrupting its own. It is often stupid, frequently wrong and not given to great leaps. It is in all this intensely human.

But democracy allows for power and non-freedom to be held in check, and for lies to be undone, and it is sustained in this by the courage of dissent and the wisdom of heresy. It is in the preservation and extension of the liberties of the people that the guarantee of the strengths and worth of democracy is to be found.

Democracy at its best is the ongoing movement of humanity towards a better world. And we see all around us that movement stalling. We see our politics broken, unresponsive to the great questions of the age, unable to name, far less address, the central challenges. It as if this great river has suddenly been halted in its path to the sea of hope.

We need to look the disease of Australia in the eye, the disease of conformity that is ill preparing us for the future. Does Australia still have the courage and largeness it once had when it pioneered the secret ballot and universal suffrage? Or will it simply become the United Arab Emirates of the West, content to roll on for a decade or two more, glossing over its fundamental problems while brown coal and fracked gas keep the country afloat? Does Australia have the desire to move into the twenty-first century, or will it continue its retreat into a past as a colonial quarry for the empires of others, its public life ever more run at the behest of large corporations, its people ever more fearful of others, its capacity for freedom and truth with each year a little more diminished?

"In reading the gospels," wrote Oscar Wilde when in Reading Gaol – and let us not forget that while there he was allowed nothing else to read – "I see the continual assertion of the imagination as the basis of all spiritual and material life, I see also that to Christ imagination was simply a form of love."

One does not have to be a Christian for this to seem a beautiful, and a very true, observation.

Yet in recent times we seem to have lived through not so much a crisis of politics as a collapse of that most human attribute, empathy, a collapse so catastrophic it sometimes appears to be a crisis of love, manifest in epidemics of loneliness and depression. It is a crisis most evident in the strange echoing cyber-caverns of the internet, where a convict in a Chinese gulag, physically exhausted, regularly beaten, may after twelve-hour days labouring in a coal mine be forced to play World of Warcraft online in a practice known as gold farming. To earn online credits for his prison bosses to trade for cash, he may find himself pitted against a far right-wing Norwegian extremist who is planning a massacre to save his idea of the world from another idea of the world that only exists online and in his mind. There is contact here, even communion. But does this world create empathy, or destroy it?

In recent years I have come to recognise the wisdom of Vasily Grossman's final novels. Perhaps our homeland is simply the people we love and who love us. Perhaps the only party of honour is the party of one. Perhaps the world advances to a better place through the countless acts of everyday goodness shown by millions of people too easily dismissed as everyday. None of this amounts to an answer to any of our problems, I know. But as Rilke said, "Live the questions." Questions lead to poetry, science, freedom. Certainties lead to Andrew Bolt's blog.

We need politics like we need sewerage, and like sewerage we should want it to work properly and well but not make too much of it, not create of it a fetish by glorifying it with daily celebrations and watching it incessantly on 24-hour TV stations. We in Australia make too much of our political leaders and their work, their failings, their strivings, their successes, and too little of ourselves. For if we take our compass from power we will inevitably arrive at despair, whereas if we take our compass from those around us we will arrive at hope.

There are so many forces in the world that divide us deeply and murderously. We cannot escape politics, history, religion, nationalism – for their sources lie as deep in our hearts as love and goodness, perhaps even deeper. In a world where the road to the new tyrannies is paved with the fear of others, we need to rediscover that we are neither alone, nor in the end that different, that what joins us is always more important than what divides us, and that the price of division is ever the obscenity of oppression.

We need once more to assert the necessity of witnessing and questioning as the greatest guarantee we can have of freedom. If I am left believing in anything, it is something very simple: that truth matters above all else. Anything that honours and guarantees the truth is not just good, but necessary. And anything, like mass conformity, that threatens the truth needs to be challenged. For the road to tyranny is never opened with a sudden coup d'état. It is a long path paved with the small cobbles of silence, lies and deceit that ends, inevitably and terribly, in horror.

In the end none of these things is ever a matter of party. Weary Dunlop most likely voted Liberal, yet it is no paradox that Tom Uren, once known as the heart of the left, said he learnt his socialism from Weary Dunlop while a POW. Tom Uren, like Weary Dunlop, didn't agree. And while a

Labor man through and through, Uren has described the Greens' Bob Brown, another man who doesn't agree, as having the blood of Mandela flowing in his veins. These are matters of character and, to use a word little heard these days, courage. More than ever, in this new age, Australians need once more to recover their voice, and that power of not agreeing with power. It's time, like Slappy Oldham, we looked after ourselves a little more, and deferred to power and its Driscolls a little less.

Like the aged Weary Dunlop we could do worse than ponder Ulysses' exhortation to freedom:

> To follow knowledge like a sinking star,
> Beyond the utmost bound of human thought.

Perhaps freedom is finally a form of love. It is unachievable, elusive, Rilke's hardest test that we are destined repeatedly to fail, the greatest challenge, the pursuit of which takes us through an endless cycle of trials and ordeals. But it is our Ithaka, and in our journeying towards it there can be no cease.

*4 September 2011*

1. Chris Berg, "Free speech: Hicks should keep his memoir profits," *The Drum*, 10 August 2011, http://www.abc.net.au/unleashed/2831862.html.

Correspondence

Nick Cater

Given the tone of Robert Manne's recent interventions in the national debate, his study of the *Australian* was unlikely to have been entirely complimentary. Nevertheless, the newspaper cooperated in the hope that Manne might return to his earlier, more scholarly style and deliver a fair and honest assessment of the masthead he anoints as the country's most important.

It is clear from his *Quarterly Essay*, however, that guilt had been predetermined: a crime had been committed, Chris Mitchell's fingerprints were all over it, and the hours of transcribed interviews Manne conducted with the *Australian's* senior staff could best be described as "helping police with their inquiries." Transparency is a wonderful thing, but opacity is cheaper, and if your appraiser has his mind made up, you're going down anyway.

For thirty years or more, Manne has distinguished himself through his rare determination to exercise his intellect in the town square. There is no sign he intends to relinquish his position as public intellectual, but with this essay his thinking has retreated further into the cloisters. He has become ever more abstract, aloof and contemptuous of his interlocutors. I mean no disrespect by suggesting that Manne needs to get out more.

To give credit where it is due, nothing from the interviews was quoted out of context. In fact nothing was quoted at all, as Manne explained to Peter van Onselen in a Sky News interview:

> I didn't find any of the things that [editor-in-chief Chris Mitchell] said sufficiently sharp or interesting to quote directly ... I internalise what people say and if other people want me to have a different style, so be it ... But I didn't feel the need to quote directly.[1]

The Media Alliance Code of Ethics instructs journalists to do their utmost to give "a fair opportunity for reply" and warns against suppressing or distorting "relevant available facts." It makes no provision for internalising a rebuttal. Manne, however, is not a journalist and loves his loaded adjectives too much to adopt the pretence of objectivity. He has become a polemicist who press-gangs the evidence into fighting for his cause. In the postmodern manner, the testing of thesis against antithesis, a method that has served scholars well since Socrates, is old hat.

It is replaced with a zealous reductionism that elbows complexity aside in the rush to nail everything to the mast.[2] Readers familiar with Mark Kurlansky's *Cod: A Biography of the Fish That Changed the World* will be acquainted with the genre. Extraneous evidence is expunged and contrary arguments ignored to reach the conclusion that, since the sun comes up each day at roughly the same time as the *Australian* lands on the lawn, Mitchell poses a threat to the planet's rotation.

I exaggerate, but only slightly. Before reading Manne's essay, I was unaware that an editorial published in March 2010 that recommended abandoning the cap-and-trade emissions proposal in the absence of international agreement was one of two that brought down Kevin Rudd. The second, on 22 June 2010, argued that the government's attempt to sell its mining tax had blown up in its face.

"Three days after this editorial, Rudd was gone," Manne writes. "At the *Australian* it was now time to gloat."

The animosity between the prime minister and his caucus is inconsequential to Manne since the friction he presumes to have existed between Rudd and Mitchell explains everything. Nor does Manne allow that the failure of the Kyoto process, the Coalition's revival under Tony Abbott or Rudd's declining popularity played any part in the events of 24 June 2010. Instead, the malevolent puppeteers of Holt Street pulled the strings that sealed Rudd's fate. In Kurlansky's world, Norway's destiny is shaped by the habits of fish. In Manne's world, Labor's destiny is shaped by the *Australian*'s perfidy.

"The unrelentingly personal and ideological campaign it waged against Rudd in the months between February and June 2010 undoubtedly helped crystallise opposition to him inside the caucus and the party. In this way, the *Australian* was a very important catalyst leading to his fall," he writes.

*Quarterly Essay* 43 is in itself a classic example of the "unrelentingly personal" narrative; Mitchell is named no fewer than 127 times and referred to as editor-in-chief on a further sixteen occasions in 115 pages. The *Australian*'s former editor Paul Whittaker is mentioned only five times, twice in reference to "riding instructions" he is supposed to have been given by News Limited's chairman and chief executive, John Hartigan. By contrast, Rupert Murdoch's enforcer, the Machiavellian, monomaniacal Mitchell, bows to no man in this implausible portrayal of a newspaper going rogue. Anyone familiar with the industrial task of publishing 310-plus editions a year, maybe 12,000 broadsheet pages and thousands more pages of magazines and supplements, will spot the plot's absurdity. In civilian life, the potential of hundreds of individuals cannot be harnessed to achieve a collective goal by bullying, as anyone who has run a medium- to large-scale business or studied a unit of industrial sociology would attest.

What weight should we give then to the allegation that the *Australian* is waging a neoliberal "ideological campaign" or the accusation that we are caught in "a truly frightful hotchpotch of ideological prejudice and intellectual muddle"? One of the main indoctrination tools, we are told, is a short daily feature called "Cut and Paste": "Everyone who reads the *Australian* knows that daily mockery of opponents is one of the most potent means by which the paper's ideological and political agenda is advanced." One imagines George Orwell's long, windowless hall, with its double row of cubicles and endless rustle of papers, where Tillotson sits with a folded newspaper on his knee, murmuring into the mouthpiece of speakwrite.

Manne should know better than to play the neoliberal card. In an interview with Terry Lane on ABC Radio National in June 2005, he told Lane: "I have never studied economics formally and found, pretty quickly when I began to argue about economic rationalism or neoliberalism, I found myself out of my depth." Years of intellectual laziness have corrupted the meaning of "neoliberalism" to such an extent that its original meaning – a set of economic principles also known as the Washington Consensus – is rarely understood. Suffice to say, it is neither an ideology nor a term the *Australian* would embrace. Good economic policy is simply policy that works, which is why mainstream economic liberalism has set the course not only for the *Australian* but for every Australian prime minister (including Rudd, despite his protestations), every US president and every British prime minister for thirty years or more. If Manne insists on wading into deep water, he might explain which of the *Australian*'s economic principles he contests: fiscal discipline, an aversion to indiscriminate subsidies, moderate marginal tax rates, market-determined interest rates, trade liberalisation or competitive exchange rates. Otherwise he should take his own good advice, as expressed to Lane: "I should stay with things that I know pretty well."

In keeping with the essay's declaratory tone, however, the reference to neoliberalism is not an invitation to debate, but an axiom for malevolence. That is the extent of the intellectual investment Manne is prepared to make. There is no need to break into a sweat since, in the world Manne inhabits, *everybody knows* that

Mitchell has debased a once-respectable newspaper by taking it for a walk in the political fringe-lands. It falls into the category of knowledge sociologist Pierre Bourdieu calls *habitus*, a matter a clan takes for granted. A more consequential study of the *Australian*'s culture, examining forty-seven years of archived back pages, would reveal that the *Australian* is, pretty much, what it always has been: a newspaper with a classical liberal outlook, a non-partisan supporter of national progress that takes a sceptical view of what *everybody knows*.[3]

The *Quarterly Essay* represents a substantial investment in the nation's cultural capital and the series is a credit to its publisher, Morry Schwartz. Its forty-third edition, however, highlights the genre's limitations. The essay's disclaimer – "Many aspects of the paper are not analysed" – admits its restricted scope. Should we, as requested, read it as "deep analysis"? Or is it, as someone rather harshly suggested, the world's longest tweet?

Manne has selected articles, from more than 1000 pages printed a month, that touch upon "ideologically sensitive questions." He would deny he is out to settle old scores, yet the "ideologically sensitive" questions follow a predictable pattern. They are causes he has staked his personal reputation upon: Aboriginal history, climate change, foreign intervention and the future of social democracy. He ups the stakes by playing the morality card: You don't have to be stupid to get caught on the wrong side of the argument, you just have to be bad. No wonder he is irritated to see his arguments challenged in the pages of the *Australian*, particularly those things that *everybody knows*. Open-minded readers would recognise this process as a robust contest of ideas. To Manne, however, each news story, commentary piece or letter contesting his view is an ideologically guided missile with his name written on it. Hence the *Australian* did not "debate" the so-called history wars, it "pursued the claims" of historian Keith Windschuttle with "partisan ferocity." Manne's argument is not with the *Australian* but pluralism.

He does acknowledge the newspaper's "vital role in alerting the general public to the breakdown of conditions of life in the remote Aboriginal communities not only in the Northern Territory but across the country." For once, however, his concern is about the voices he claims we did not publish, rather than the ones we did: "the neglect of such voices itself represents a kind of distortion." Manne cannot have expected this statement to go unchallenged. If he cares for a list of the Aboriginal commentators we have published, I will arrange for our library to send him one, but he knows as well as I do that both in number and in column centimetres they amount to many more than the total of every other mainstream newspaper combined.

Manne acknowledges the *Australian*'s opinion page was "reasonably balanced" in the period leading up to the invasion of Iraq; however, he considers the "tone" of editorials and commentary by foreign editor Greg Sheridan "significant and revealing." What do they reveal and why is it significant since the *Australian*'s support of the invasion of Iraq is a matter of public record? It is "significant and revealing" perhaps that Manne quotes Sheridan, but not Paul Kelly, who under the headline "The Hapless Persuader" highlighted the weaknesses in John Howard's case for war in March 2003. "He has failed to mount a persuasive argument that a war to disarm Iraq is an imperative now when the risks are so vast and the national interest could be prejudiced," Kelly wrote. Manne was right the first time; the opinion pages were "reasonably balanced." Once more, his quarrel with the *Australian*'s editorial position can be summarised in four words: He does not agree.

Manne acknowledges that "the claim the *Australian* acted as an apologist for the Howard government is wrong" but takes us to task for our unreliability as a cheerleader for Kevin Rudd. The professor is affronted that the *Australian* contested Rudd's February 2009 essay, which blamed extreme capitalism and neoliberal governments for the global financial crisis. Manne, who does not pretend to be an economist, cannot allow that the *Australian* does not, and never has, advocated a crude, unrestrained free market. The Rudd government's decisive first-round response to the credit crisis earned the prime minister our Australian of the Year Award.

Manne simply asserts as fact that our objections to government economic intervention on issues such as home insulation and the school-building program were ideological. Editorials from that period make it abundantly clear that our concern stemmed from prudence, not principle. Manne agrees that the criticism of the home-insulation scheme was "neither ephemeral nor trivial." He disagrees, however, with our assessment that the school-building program was expensive, rushed and wasteful. In our view, wasting up to 12 per cent of the $16 billion spent on the Building the Education Revolution is a problem. Manne disagrees; end of story.

Once the neoliberal explanation has collapsed in a heap, Manne is obliged to call upon his infinite capacity for astonishment. He is "astonished" that the *Australian*'s editor-in-chief should ask journalism academic Julie Posetti to retract allegations of unprofessional behaviour; Mitchell publishes an "astonishing" editorial criticising the ABC's *Media Watch*; "astonishingly," a story was published on the nuclear industry. Yet Manne still has not learned that the *Australian* is not his kind of newspaper and laments the "astonishing" imbalance of its opinion columns, its "astonishing" pro-mining bias and the "astonishing" attempt to wreck the Labor–Greens alliance.

When the astonished Manne draws breath, the moraliser takes over, condemning a "disgraceful saga of protracted character assassination" against fellow academic Larissa Behrendt. The *Australian's* coverage of the Behrendt–Bess Price tweet controversy is "truly foul" (he offers no such condemnation of Behrendt's tweet). Behrendt is the chief victim of injustice and the *Australian* is her persecutor, further evidence of a "bullying culture" driven by the "bullying behaviour" of the editor-in-chief. He cannot allow that the *Australian* merely contributes to debate, or give any weight to our editorial judgment:

> The Twitter exchanges reveal the split between urban and remote Aboriginal leaders over Canberra's intervention in dysfunctional communities. Behrendt's comments are made against the background of a bitter struggle between these two groups for power and the authority to speak for Aborigines. Behrendt and those who joined her on Twitter oppose the intervention but that is really a proxy for a fight over turf, resources and the direction of indigenous politics.[4]

On climate change, the charges are grave: the paper is accused of undermining "the central values of the Enlightenment, Science and Reason." The motive? "Emphasis should be placed on the role of ideology in rationalising the defence of mining interests and helping to subvert reason." The passive sentence construction betrays an author unsure of his ground.

To arrive at his findings, Manne adopts a "complex methodology": he downloads every climate change article between January 2004 and April 2011, including "Cut and Paste" columns, and reads them. He tells us that 180 pass muster and are therefore deemed to be "favourable to climate change." Some 700 are deemed "unfavourable," thus demonstrating that the *Australian* favoured the unfavourable by a ratio of four to one. Later, Manne ups the charges to a ratio of ten to one, for reasons that are not properly explained. Peer review is impossible since Manne does not share his raw data and the judgments are puzzling. Regular contributor Bjorn Lomborg, for example, who is convinced that man is contributing to climate change but is unconvinced that the Kyoto process can solve the problem, is judged a denialist. Yet Frank Furedi, condemned by Manne in August 2008 as "a leading climate change denier," disappears from the list of Most Wanted Repudiators.

Broadly, Manne complains of "dozens" of "denialist" articles by twenty named "denialists" over seven years. In his later contribution to the *Weekend Australian*, he raises that to "scores." By my count, the number of articles written by the accused is: Bob Carter (6), Michael Asten (2), Lord Monckton (2), Ian Plimer (4), Jennifer Marohasy (4), Garth Paltridge (1), Dennis Jensen (2), John Christy (reprinted from the *Wall Street Journal*) (1), David Evans (3), David Bellamy (1), Nigel Calder (reprinted from the *Sunday Times*) (1). There were joint-bylined pieces: one by Richard Lindzen, John Roskam and Ullrich Fichtner and another by Bob Carter, David Evans, Stewart Franks and Bill Kininmonth. That's a total of twenty-nine, which means Manne was right about "dozens," though not about "scores." Roughly speaking, it represents one "denialist" argument for every 350 published opinion pieces.

The final pages of *Bad News* should be read as a tragedy: an academic who once published respectable books finds himself down on his luck in Conspiracy Corner, the two-dollar shop for irrefutable arguments. News Limited "represents a threat to the flourishing of an open democratic culture." Rupert Murdoch is "a highly political individual with a powerful set of ideological beliefs." His company has "a stranglehold over the daily press" and "must be challenged" since it has the capacity "to influence … the vast majority of less engaged citizens." A plot "to do something" about the Labor–Greens alliance, hatched at a secret meeting of editors and senior journalists in the US earlier this year, "poses a real and present danger to democracy." The *Australian* hides behind "the guise of a traditional broadsheet newspaper" but is in fact the "national enforcer of those values that lie at the heart of the Murdoch empire: market fundamentalism and the beneficence of American global hegemony." It relies on a "hidden financial subsidy from the global empire" and is used by Murdoch as a "means for influencing politics and commerce in the country of his birth." Not everyone at the *Australian* is part of the conspiracy; there are some real journalists, but Manne "will not name them for fear of doing them harm."

The narrative carries the hallmarks of what David Aaronovitch calls "voodoo history," a genre of conspiracism in vogue in Western countries in which books alleging secret plots appear on the shelves alongside scholarly works by noted academics.[5] There is a threat to *our very way of life* (democracy) by two or more *sinister individuals* (Murdoch and Mitchell) driven by *a dark force* (neoliberalism) and with the power to manipulate not just their immediate underlings (the staff) and the masses (less engaged citizens) but the people who run the country (the Labor government). A secret meeting (the Carmel gathering) is held to hatch the plot (the overthrow of the Gillard government).

Sadly, if I am right, then this reply to Manne's essay, the metres of newsprint columns devoted to answering his allegations and the interviews given to assist his research have been a waste of time. Conspiracy theories are resistant to reason, since their authors are skilled at blocking contrary arguments. Worse still, once the theorist has promoted himself as *the one who understands the true order of things*, he is obliged to keep the narrative alive for fear of suffering catastrophic reputational damage. Tucked away in the notes at the end of his essay, Manne admits he was approached to write for the *Australian* in February 2009. In conversation with me on stage at the Byron Bay Writers' Festival, Manne further acknowledged my open invitation to contribute whenever he wants on whatever subject he wants.

> Nick came down to Melbourne, he offered me space in the paper. Our personal relations, I want to say this publicly, he's been courteous, he's published anything that I suggested. I don't want to write for the *Australian* because I want the *Australian* to change and I don't want to give it legitimacy by writing for it.[6]

Had Manne written for us on climate change once every six weeks, his contributions would have outnumbered those by the listed "deniers" by two to one. His refusal to do so is one thing, but his subsequent complaint that our coverage lacks balance draws me reluctantly to the conclusion that Manne is not acting in good faith.

The *Australian's* first edition – on 15 July 1964 – threw the newspaper open to "thousands of friends who, as the thinking men and women of Australia, will have a profound influence on the future. You are welcome to this company of progress." It saddens me that Manne has rejected the invitation with such bad grace.

Nick Cater

1 *The Showdown*, Sky News Australia, 20 September 2011.

2 See Daniel Dennett, *Darwin's Dangerous Idea*, Simon & Schuster, New York, 1995.

3 See Philip Roth's soliloquy on what everyone knows in *The Human Stain*, Vintage, London, 2001.

4 *The Australian*, 15 April 2011.

5 David Aaronovitch, *Voodoo Histories*, Jonathan Cape, London, 2009.

6 Manne in conversation with the author, Byron Bay Writers' Festival, 7 August 2011.

Jay Rosen

In what is perhaps the most important sentence in his essay, Robert Manne writes, "With regard to the problem of the *Australian*, I can think only of one possible solution: courageous external and internal criticism." He has given us that. And I am grateful for it.

Though I don't live in Australia, and do not have the same stake in its public culture, I am grateful to Manne because the worldwide Murdoch empire cries out for learned criticism. I don't think we understand it very well. In my contribution to this forum, I want to sketch a brief theory of News Corp., which will assist in interpreting Manne's exemplary work.

News Corp. is a huge company, but it is not a normal company. However, it does not *know* that it's not a normal company. In fact, it denies this observation. In this sense denial is constitutive of the company and its culture. To work there, you have to share in this pervasive atmosphere of denial. And that's how many of the things Manne details so well can happen.

For example: the *Australian* is a force for climate change denialism. But it does not know this about itself. Outsiders do know it, and they regularly point it out. The *Australian* reacts not by defending its actual stance on climate change but by trying to destroy those who accurately perceive it. The attempt at destruction is typically rhetorical but sometimes other methods are used, like threatening a lawsuit. The impression given is of a bully or thug. But that's really an after-effect of denial. Denial, I think, is the key to understanding the company.

News Corp., as I said, is not a normal company. One thing this means was pointed out by *Fortune* magazine columnist Geoff Colvin:

> Some people aren't at all surprised by the unending scandal at Rupert Murdoch's News Corp. They are the investors, insurers, lawyers, and others who had read the "Governance Analysis" report on the

company from The Corporate Library, a research firm. The firm grades companies' governance from A to F, and for the past six years News Corp. has received an F – "only because there is no lower grade," says Nell Minow, who co-founded The Corporate Library in 1999 on the premise that governance "can be rated like bonds, from triple-A to junk." News Corp.'s overall risk, says the prophetic report: "very high." Risk of class-action securities litigation: "very high." Scandal-related lawsuits are already piling up.[1]

Why does this matter to the story Manne has to tell? Colvin tells us:

> The effects are insidious and more far reaching than you might imagine. "It creates a culture with no accountability," says Charles Elson, director of the University of Delaware's John L. Weinberg Center for Corporate Governance. In companies where directors are genuinely subject to the shareholders' will, CEOs get fired; BP's board fired Tony Hayward last year, for example, and Hewlett-Packard's board fired Mark Hurd. The message cascades down through the organization: Bad behavior gets you fired here. But at companies where the CEO can fire the board, a different message cascades down: We don't answer to the shareholders, we answer to just one person. It's the rule of man, not the rule of law.

In other words, Chris Mitchell, editor of the *Australian*, is in every way a creature of News Corp. He has within him its DNA. Case in point, ably discussed by Manne: it is simply unimaginable that a top editor at a major American newspaper could threaten a journalism lecturer with a lawsuit for accurately reporting the contents of a speech in a public forum, simply because the speech reflected poorly on the editor's judgment. The contradiction between personal vanity and the professional imperative to defend honest reporting would be too great to sustain. But not within News Corp. Mitchell knew his threat against journalism educator Julie Posetti would appeal to Murdoch's self-image as a brawler, and that's all that counts.

Watching this action from afar (Posetti is a friend of mine), I found it especially appalling to observe how rank-and-file journalists at the *Australian* fell in line behind Mitchell's action, lashing out at critics when the reality was that the threatened lawsuit humiliated them, because it demonstrated publicly what I just said: personal vanity can trump an editor's professional imperative to defend the work of honest reporting. The more trouble plaintiffs can make for journalists who are just trying to report what happened, the weaker the press in any country. Everyone who works in a professional newsroom knows this. But denial came easily to the staff of the *Australian*, because it is part of the atmosphere they breathe.

The American journalist Carl Bernstein remarks on how an atmosphere like that is created:

> As anyone in the business will tell you, the standards and culture of a journalistic institution are set from the top down, by its owner, publisher, and top editors. Reporters and editors do not routinely break the law, bribe policemen, wiretap, and generally conduct themselves like thugs unless it is a matter of recognized and understood policy. Private detectives and phone hackers do not become the primary sources of a newspaper's information without the tacit knowledge and approval of the people at the top, all the more so in the case of newspapers owned by Rupert Murdoch, according to those who know him best.
>
> As one of his former top executives – once a close aide – told me, "This scandal and all its implications could not have happened anywhere else. Only in Murdoch's orbit. The hacking at *News of the World* was done on an industrial scale. More than anyone, Murdoch invented and established this culture in the newsroom, where you do whatever it takes to get the story, take no prisoners, destroy the competition, and the end will justify the means.[2]

The *Australian* conducts itself like a thug, in all the ways Robert Manne shows us, because it understands that this is News Corp. policy. But the policy isn't a formal one; that's for fools, which is why News Corp. gets an F in corporate governance. Rather, thuggishness is part of the culture of the company. Let me give you another example, a moment when Rupert Murdoch himself humiliated the newspaper he is proudest to own: the *Wall Street Journal*. During the worst days of the phone hacking revelations in the summer of 2011, Murdoch rang up a reporter working on a story about the company and boasted about its performance. News Corp. has handled the crisis "extremely well in every way possible," he said, making just "minor mistakes."[3]

This was ludicrous, the type of claim that a seasoned *Wall Street Journal* reporter couldn't possibly accept with a straight face. In fact, events made an absolute mockery of it. The next day the executive in charge (Rebekah Brooks) resigned.

Two days later Brooks was arrested. I guess she hadn't handled the crisis "extremely well in every way possible." Then Les Hinton, Murdoch's closest aide, resigned in hopes of reversing the tide of defeats. Your top executives don't quit for what Murdoch called "minor mistakes," and yet his transparently dishonest statement ran unchallenged in the *Wall Street Journal*. The boss was in denial. The reporter who interviewed him and the editors who approved the story knew their role: to keep the culture of denial going.

Here, then, is my brief theory of News Corp., taken from the essay on this topic that I published in the *Guardian* during the most intense period of revelations in the phone hacking scandal that is still underway in the United Kingdom.[4]

News Corp. is not a news company at all but a global media empire that employs its newspapers – and in the United States, Fox News – as a lobbying arm and intimidation machine. The logic of holding these "press" properties is to wield influence on behalf of the (much bigger and more profitable) media business and also to satisfy Murdoch's own power urges or, in the case of Australia, his patrimonial legends.

But this fact, which is fairly obvious to outside observers like Manne, is actually concealed from the company by its own culture. This, then, is the source of the river of denial that runs through News Corp. Fox News and newspapers like the *Australian* are understood by most who work there as "normal" news organisations. But they are not. What makes them different is not that they have a more conservative take on the world – that's the fiction in which opponents and supporters join – but rather: news is not their first business. Wielding influence and intimidating people is. Scaring politicians into going along. Building up an atmosphere of fear and paranoia, which then admits Rupert in through the back door of 10 Downing Street.

But none of these facts can be admitted into company psychology, because the flag that its news properties fly, the legend on the licence, doesn't say "lobbying arm of the Murdoch empire" or "influence machine." It says "First Amendment" or "journalism" or "public service" or "news and information." In this sense the company is built on a lie, but a necessary lie to preserve certain fictions that matter to Murdoch and his heirs.

Strangely, I do not think that News Corp. people like Rebekah Brooks, James Murdoch and Chris Mitchell are being insincere when they pledge allegiance to the values of serious journalism. On the contrary, they believe that this is what their newspapers are all about. And this is the sense in which denial is constitutive of the company, a built-in feature that cannot be acknowledged by any of the major players because self-annihilation would be the result.

Robert Manne is right that the only defensible reply to the problem of the *Australian* is vigorous criticism. Action by government would damage the principle of a free press. What I am trying to point out is that such criticism not only has to point out bad behaviour and irrationalism. It also has to somehow surmount the culture of denial that has helped to create the *Australian*. A culture of denial would be a troublesome beast in any company. In a newspaper it is menacing, and this is why the *Australian* is such a menace.

Jay Rosen

1 Geoff Colvin, "The trembling at News Corp. has only begun," *CNN Money*, 19 July 2011, http://management.fortune.cnn.com/2011/07/19/the-trembling-at-news-corp-has-only-begun.

2 Carl Bernstein, "Murdoch's Watergate?" *Newsweek*, 9 July 2011, http://www.thedailybeast.com/newsweek/2011/07/10/murdoch-s-watergate.html.

3 Bruce Orwell, "In interview, Murdoch defends News Corp.," *The Wall Street Journal*, 14 July 2011, http://online.wsj.com/article/SB10001424052702304521304576446261304709284.html.

4 Jay Rosen, "Phone hacking crisis shows News Corp. is no ordinary news company," *The Guardian*, 19 July 2011, http://www.guardian.co.uk/commentisfree/cifamerica/2011/jul/19/rupert-murdoch-phone-hacking.

Tim Flannery

Robert Manne's *Bad News* documents, among other things, evidence of bias in the *Australian*'s reporting of climate change. Here I wish to add a note explaining how right-wing media in Australia can manufacture news to suit their bias. The example concerns my role as chief climate commissioner and the issue of sea-level rise, as reported by 2GB, the *Daily Telegraph* and the *Australian* in July and August 2011.

On 28 July, just six days after a right-wing Norwegian terrorist killed seventy-six people whose views he disagreed with, the 2GB opinionist Ray Hadley announced on air that he'd received a call from a man identified as "David." He claimed to be a neighbour of mine who lived on the Hawkesbury River. The 2GB website states: "'David calls in to Ray Hadley to confirm' that Tim Flannery owns a waterfront home."

David described exactly where my home is and gave a (somewhat misleading) description of it. This breach of privacy had implications for my security and that of my family. Led by the far right, the climate debate in Australia had by this time become hysterical. Several scientists associated with the climate commission had received death threats, and the commission had found it necessary to beef up security at its public events.

Prompted by David's "revelation" that I lived by the water, Hadley inferred that my home was vulnerable to inundation by sea-level rise. This inference is false (the house is in fact located well above the 1.1 metre mark above sea level), but was taken by Hadley as evidence of hypocrisy on my part. Urged on by Hadley, David then implied that I broke the law by speeding through 4-knot zones and had a highly polluting outboard motor (again false).

Hadley's justification for running the interview was clearly my supposed hypocrisy: "It's terribly important that people know about these people," he said by way of justifying putting the conversation to air.

Several weeks later I had established the identity of the caller David and the location of his home. On the afternoon of Sunday 21 August I visited him and questioned him about his call to 2GB. His stammering voice was so unlike the smart-aleck tone I'd heard on the radio that at first I thought I had the wrong person. But he soon admitted that he knew Ray Hadley. In fact, he said, he worked for him.

David then stated emphatically that he had not called Ray Hadley at all. Instead Hadley had asked him to appear on the show, and had called him. David said that Hadley had sought him out after learning that I lived nearby. The story, and all of the supposed "facts" that David was to raise during the interview, had, according to David, been assembled beforehand by Hadley and his team. Notes taken by my wife as I spoke with David reveal the true motive for the interview. David stated: "You're on the other side of the fence [regarding climate change] … they hate you … they're out to get you."

What a listener might have taken to be a neighbour dobbing in a hypocrite and law-breaker was in fact a slander based on a completely manufactured story.

Hadley's piece was soon taken up and embroidered by other right-wing opinionists. Andrew Bolt ran a blog which further invaded my privacy by adding details of our home's value at purchase and its mortgage status, and needlessly invaded the privacy of my wife, who is not a public figure and has no role in the climate debate. He also included a photograph, taken from Google Earth, of our home.

The publication of the photograph of my home concerned the security officer at the climate commission. She briefed the Australian Federal Police on the situation. Sydney's *Daily Telegraph* ran an article drawn from Bolt's blog, but the photograph was removed from the paper's website following a call from a QC to the newspaper's editor.

I then began receiving phone calls from journalists at the *Australian* requesting an interview for stories relating to sea-level rise. Leigh Dayton called me several times, asking for an interview for a serious article on sea-level rise that she said she was researching. In her last call she stated that she intended to do some serious journalism, and that "I wouldn't lie to you, Tim."

On Friday 5 August Ean Higgins of the *Australian* called the climate commission requesting an interview with me for "an intelligent article" he was writing on climate change. As he had informed the commission's publicist that he was interviewing a number of experts for the story, it was recommended that I take his call.

I gave Higgins a thirty-minute interview on the work of the climate commission in relation to sea-level rise, virtually none of which made it into his story.

Instead Higgins wrote a re-hash of the Hadley story. As with Hadley, the rationale was that I had been hypocritical in purchasing property vulnerable to inundation by sea-level rise. I had pointed out to Higgins during the interview that this was untrue — a fact that did nothing to blunt his compulsion to publish.

Higgins could not resist embroidering his story, adding a comment from the blogger "Old Salt." But Higgins went further than the blogger, stating in the article that I was swindling the elderly out of their properties by frightening them into selling up with talk of rising sea levels. Accusations of hypocrisy had now become malicious libel, and I decided to challenge the *Australian* legally.

My action against the *Australian* resulted in the withdrawal of the article and the publication on the following Saturday of a page-three apology, as well as payment of all legal expenses, amounting to $5000.

The experience has taught me several things about the hate media in Australia. First, as they seek to slur those they hate, they do not hesitate to manufacture a story if one does not exist. Second, as the story is picked up by other opinionists, they are prone to weave ever more scandalous fictional tidbits from the blogosphere into the story. Third, in their efforts to obtain an interview some journalists will lie and ignore the truth when it's inconvenient to them.

Perhaps most dismayingly, the *Australian* has chosen, to its eternal discredit, to publish such detritus. If even our national newspaper can't keep its nose clean, strict rules around media misconduct can't come soon enough.

Tim Flannery

Mark Latham

One of the curious distinctions in Australian public life is the difference between journalism and broadcasting. Most notably, Ray Hadley uses it to justify his blurting, opinionated style on Sydney radio's 2GB. To paraphrase his argument, journalists are expected to be impartial and thorough in their work. It is acceptable, by contrast, for broadcasters to be politically biased and extemporaneous in their pronouncements. As is clear from Hadley's show, there is no compulsion to check the facts or ensure the other side of politics receives fair treatment. There is a lot of broadcasting at 2GB, the equivalent of a continuous-play advertisement for the Liberal Party. It is a campaigning style of media, with a premium on ideology and electoral impact.

In light of Robert Manne's detailed essay on the prejudices of the *Australian* newspaper, it is timely to reassess its status. It can no longer be considered a newspaper in the traditional, journalistic sense. Rather, under Chris Mitchell's guidance, the *Australian* has become a printed version of radio broadcasting, albeit with greater finesse and substance than the likes of Hadley. The key test in distinguishing between journalism and broadcasting is this: does a media outlet seek primarily to inform its readers about news and current affairs or is its core purpose to change the nature of politics and public policy? Manne has demonstrated that the latter is the dominant tendency at the *Australian*. It has evolved into what might be thought of as a campaign-paper.

As some readers might already appreciate, I am not part of a cheer squad for the integrity of modern Australian journalism. So please do not regard me as puritanical about the *Australian* or as a defender of the old way of doing things. In a free society all manner of opinion and ideology should flourish in the media. Even though party politics is a much diminished institution in Western society, there is still something admirable in the publication of a well-constructed ideological position, whether left or right, statist or libertarian. The issue is more one of transparency than content. I would have far greater respect for Mitchell and his paper if he was upfront and honest about its agenda.

Instead of hiding behind the facade of conventional journalistic values, the *Australian* needs to be more comfortable in its own skin. It should simply declare itself to be a campaign-paper, a bold and brazen agenda-setter for free enterprise and libertarian ideology. It should drop its "Heart of the Nation" marketing claptrap and attention to socio-demographic pie-charts and proudly declare on its masthead: "The *Australian* – A Right-Wing Voice for a Right-Wing Nation." Its many critics, Manne and myself included, would then have little to complain about. Everyone would know where the paper stood and the silly game of charades could end.

Under the current arrangements, perversely enough, a ratbag like Hadley has greater legitimacy than the *Australian*. Sure, members of the political class recognise Hadley's limitations, his lack of education and simplistic outlook on issues, but they at least know where he stands – he has the great Australian virtue of being fair dinkum. This is what the *Australian* lacks, the integrity of being itself and coming at its political opponents front-on. In its current guise, it is throwing cowardly king-hits, so much so that a mild, civilised figure such as Bob Brown has labelled it "the hate media."

The clever thing about the Manne essay is its methodology. Usually accusations of bias are subjective – sweeping claims which are readily dismissed by the media outlet in question. Professor Manne avoids this problem by presenting seven case studies, a forensic assessment of the *Australian*'s campaigning style in areas as diverse as Aboriginal history, the invasion of Iraq, climate change and the role of the Australian Greens. Regular readers of the paper could extend this list many times over. Just as political parties have a standard set of campaign tools, the *Australian*'s kitbag is well established. It campaigns through a formulaic process of issue repetition, prominent story placement, reporter bias and personal sniping in its "Cut and Paste" section, each aspect of its work coordinated by a relentless editorial team determined to prove its political point.

Much has been made of the *Australian*'s inflated response to Manne's essay. For the paper's critics, this reinforced an important truth: Mitchell and his senior staff are so self-absorbed that they perceive an attack on their credibility to be a national event. A conventional newspaper would have ignored Manne and got on with reporting the news of the day. Instead the *Australian* launched a campaign against him, an exercise so typically self-indulgent it was comical.

The rebuttal that caught my eye was Mitchell's foray into opinion writing. It was, quite frankly, an embarrassment to the paper he leads. Undergraduate and disjointed in style, it ended with the strange assertion that:

deep-Green Rob Manne is not interested in facts. He wants a media inquiry because, after all his protestations to the contrary during the Howard years, he really is a keen supporter of "silencing dissent," to quote his mentor Clive Hamilton's book. To paraphrase another high-profile commentator on media, I say to editors at Fairfax and the ABC, don't publish crap just because it's written by Rob Manne. Can't be that hard.

More than any other person I know in Australian public life, Professor Manne values debate and dissent. Indeed, his intellectual journey – oscillating left, right, left – has been based on the ideals (and very often, the personal courage) of dissent. I am afraid the crap, so-called, belongs to Mitchell.

This is evident elsewhere in his article, notably in his claim that Manne's "entire critique is an ad hominem attack by a man so unburdened by knowledge that as a professor of politics he has not even grasped his own federal government's political position." But how much does Mitchell know about Australian politics? Not a great deal, evidently, given his belief that in June 2010 "Mr Rudd could not muster a single supporter for a spill and had to walk away from the prime ministership."

The spill for the Labor leadership, in fact, was instigated by Julia Gillard and her supporters within caucus. As Politics 101 students would realise, it is not possible for a Labor leader to move against himself. By definition, the spill comes from the leadership challenger. In any case, Rudd had more than a single vote in the leadership contest. At a minimum, Anthony Albanese's hard-left sub-faction was still backing him, as was the veteran senator John Faulkner. Rudd failed to contest the caucus ballot because he did not have *enough* votes, not because he had just one vote. If Mitchell does not understand these basic elements of national politics then he should not be editing a national newspaper. Even by the lax standards of the Murdoch media, he has risen beyond his level of competence.

Temperamentally Manne is an optimist, a believer in the redemptive power of reason. He wrote his critique hoping to improve the quality and fairness of the *Australian*. Whereas others simply despair about the situation, he sees a way forward through "courageous external and internal criticism." Journalists from all media houses, including News Limited, should rise up and force improvements. "If such people acted together to make their opinions known," Manne believes, "it is not impossible that change might come."

This optimism, however, is misplaced. Short of a *News of the World*–type scandal, nothing anyone says or does will change the way in which the *Australian* operates.

Its culture of introspective arrogance is too deeply embedded. It wears criticism, no matter how logical and penetrating, as a badge of honour, as a rationale for continuous campaigning and fighting off its "enemies." In the parallel universe in which it dwells, it sees itself as permanently under siege from the bourgeois left at the ABC and Fairfax, and inside Labor and the Greens.

In any case, the likelihood of Australia's journalistic class finding the courage to criticise the *Australian* is minimal. By its nature, journalism is a calling for reluctant, timid souls – people who watch others doing things of note and then write about it in the isolated safety of their press-gallery cubicles. If they were strong, gutsy personalities they would be politicians themselves, or entrepreneurs or entertainers – people who are doers in life, not voyeurs.

I witnessed a stunning example of this phenomenon during the last federal election campaign. On the day of Tony Abbott's policy speech in Brisbane, working as a novelty reporter for Channel Nine's *60 Minutes*, I asked Laurie Oakes if I could interview him about his role in the campaign (as the recipient of the notorious Rudd leaks which damaged Labor so badly). His response was to stand up and run away. In lacking the courage to answer my questions, how could someone like Oakes, a so-called leader of the press gallery, take on News Limited's flagship paper? He can talk down the lens of a TV camera (again in the isolated safety of his press-gallery cubicle) but anything else is beyond him.

A particular point of frustration for Manne is "the strange passivity of the two mainstream rivals of the *Australian*, the Fairfax press and the ABC, even in the face of a constant barrage of criticism and lampooning." The most obvious reason why journalists at Fairfax and the ABC have been reluctant to take up this fight is employability. News Limited owns 70 per cent of the newspaper market in Australia, meaning that most journalists at some time in their careers are likely to work for it. Plus News has a reputation for dealing with its critics brutally. This is the power of the bully through the ages, the ability to intimidate through the notoriety of fear. As the actress Sienna Miller, one of the abiding heroes of the phone-hacking scandal in Britain, said recently: "Everyone was scared of Rupert Murdoch, even governments. People are terrified for their reputations. They want the press on their side."

The good news is Robert Manne has stood up to the bullies in this country with a well-researched and argued essay. The bad news is that reform of the *Australian* is highly unlikely.

Mark Latham

Jack Waterford

If there were any great doubt about the general truth of what Robert Manne said about the *Australian*, about its guiding genius, Chris Mitchell, and, more generally, about News Limited newspapers, it would have to have disappeared in the month after publication. First there was a dead silence, then, as the most self-conscious, self-referential and boastful media organisation in Australia realised that it was once again at the centre of attention, it decided to join the conversation.

Editor-at-large Paul Kelly was first to examine the Australian's conscience, and the first to decide that it was without sin; moreover that Manne's frame of reference was as the compulsive censor who wanted to control the debate. A battery of artillery, including most of the newspaper's pet polemicists, was invited to opine on different chapters of the essay, to draw attention to nice things said along the way and to ridicule as false, ridiculous, exaggerated or special pleading anything which tended to show the newspaper as one given to abuse of its pulpit, abuse from its pulpit or of failure to understand the difference between the pulpit and the noticeboard. All were entirely predictable, though I do not think for a second that the views were channelled from Rupert, sculpted by Mitchell or written so as to curry favour with either. There are good minds in the executive soviet, but they do tend to think alike, which is one of the reasons they are there. Amusingly, later the *Australian* published a critical book review of the Manne essay, written by Professor Matthew Ricketson of the University of Canberra, who was appointed – after he had written the review – to the federal government's inquiry into the media. Ricketson found most of Manne's criticisms persuasive – as I do (if, like Ricketson, with some reservations) – but I do not expect that Mitchell would have hesitated for a second before publishing it. A week later, indeed, he published Manne's response to the criticisms made of him.

In the meantime, the seas through which News Limited was sailing had become a little more rough. The scandals in Britain were not going away – indeed were getting worse and closer to Murdoch central and were both emboldening once fawning politicians and inspiring revenge attacks from once loyal, now ditched, retainers. Calls for an inquiry into the Australian media were being made by both the Greens and from within caucus, and, ultimately, cabinet agreed. Although ministers insisted that it was not an inquiry into competition or bias – but to be focused on regulation – the clear intended target of the inquiry was News Limited, its dominance of the newspaper market and its apparent agenda of regime change. The Manne essay might have served as the text.

It also emerged that News Limited is itself concerned about its reputation in the market and is contemplating rebranding itself, not least as the company which reflects and fights for "the values of middle Australia." It is no coincidence that the line of attack on Manne and his essay – including in rather rambling editorials on other issues such as the Federal Court verdict against Andrew Bolt – is that Manne and those who agree with him are of the intellectual elite, desperately out of touch with middle Australia and determined to feed it only politically correct propaganda. In this narrative, as the *Australian* put it on 1 October, there is "a contest between those who believe democracy is a top-down exercise in which the views of elites take preference over the rest, and those who believe democracy springs from the people. In this debate between common sense and intellectual opinion, our inclination is to defer to the wisdom of the crowd."

It's a self-serving argument, if not necessarily for the reasons that Manne suggests, and if not necessarily one that places News Limited wrongly. I have always argued that the actual power of the Murdoch press to influence events is much exaggerated, not least by News Limited itself. Politicians have quavered, and a disturbing number, of all sides and in many countries, have fawned on Murdoch in case he has this power, but Murdoch protects his reputation for appearing to anoint successful politicians by closely following opinion polls – in effect the mob. On such occasions, most Murdoch newspapers move as one and, generally, in a completely and obviously over-the-top manner. All the more so that when the expected result occurs, Murdoch and his newspapers can claim, first, credit, then a place at the table, and, often, favours. In this respect the *Australian*, or the Sydney *Telegraph*, or the *New York Post*, are not much different from the *Sun* of London, or, once, the *News of the World*. And all too often, even the van-quished see the result as a confirmation of Murdoch's power and redouble their efforts to win back his favour, not least by giving his most critical journalists leaks and privileged access to stories. How much more savagely do these poodles finally bite back if they sense the empire is toppling, or if they judge themselves as having nothing to lose.

It is sometimes forgotten that newspapers did not emerge in history as sheer journals of news or accurate information. They came first from pamphleteers and polemicists, often lying and quarrelsome ones. Notions about the separation of news and comment or some sacred trust about the facts came relatively late. And it was relatively late for great newspaper proprietors – the Fairfaxes, for example – to drop the idea that the ownership of newspapers must be accompanied by moral purpose, though not so much fidelity to the truth as to "true" and worthy ideas (the facts might be ignored if they did not suit the general sermon). Murdoch's father – a clergyman's son – could invest his calling with moral phrases, but it is rather to Murdoch's credit that he has generally eschewed the notion of God the Father being at his side, preferring instead some anarchistic glee at seeing the mighty humbled, while making money and frightening the hell out of politicians.

I like the *Australian*, even if I am, like Manne, exasperated by it. It has often excellent journalism, and the instinct for hard – even inconvenient – news is so powerful in the News Limited culture that its editors can point, rightly, to many occasions when they have caused deep and lasting embarrassment to their pet ministers, pet causes or standing crusades. It is seriously interested in policy and on a host of issues has more journalists – and sometimes commentators – of more breadth and depth than most serious newspapers. Although most of its pet commentators are ideological warriors who are entirely predictable and often (at least by tenth repetition) quite tedious, there are arguments and essays of breadth, depth and material to surprise, sometimes even to delight. On a host of issues – if not on its obvious pet peeves – the *Australian* is the least predictable and the best read in the nation. Manne himself says this. A good deal of this is owed to Mitchell and his restless search for news, for angles, for ideas and for causes. Not a little is added by a coterie of backbenchers keen for an angle in a story – in part because the *Australian* is configured on the basis that it is a national newspaper, and that it cannot be (though it often is) the first, or primary, news source in most of the houses to which it comes. What it does, accordingly, is seek to have not only the most analysis and interpretation, but the highest overview and the writers of best reputation.

One sees this particularly in Canberra. Although there is only one locally produced newspaper (since, in 1967, the *Australian* failed to supplant the *Canberra Times*), the people of Canberra have easy access to at least ten daily newspapers, and a significant proportion of the market take at least three – generally the

*Canberra Times* and some of the Sydney papers, the *Australian* or the *Financial Review*. The evidence suggests not that readers find individual newspapers unsatisfactory, but that they are eager for a wide array of different takes, different interpretations and different analysis. Increasingly, too, they are loyal not only to particular newspapers, but also to particular journalists and name writers. (And – opportunely as much as ominously – the most avid readers are also the highest consumers of news and comment on the internet, the radio and television.)

In such an environment – remembering too that Canberra has a "political class" probably several times larger than in any other marketplace – Manne need not worry that readers are deceived by the obvious bias, bile or agendas of Chris Mitchell. Nationwide, indeed, more than half of the readers of the *Australian* are members of the managerial and professional classes – people not at the top of the list of those likely to fall for mere propaganda. Indeed, there might be a case for arguing (as some, even inside News Limited, have done) that the *Australian* would be more influential, and, possibly, better read if it were not so hectoring, so bullying, so single-minded and so bloody infuriating.

That said, the worst mistake the next editor-in-chief of the *Australian* could make would be to attempt to mimic the *New York Times* or some other impersonal journal of record. If Murdoch has exported some poor journalism, he has also understood that newspapers must have personality, character and a point of view – one that generally comes from the editor. Newspapers are, after all, read only by volunteers, and a decreasing number at that. Those volunteers are, generally, those of our citizens who are most interested in the world and most likely to be able to apply reason to their judgments of events. Both Mitchell and Manne underestimate their readers. But Mitchell is wrong, I think, in saying that Manne simply wants his thoughts imposed on readers. He is even more wrong in thinking that readers who agree with him require constant reinforcement, while those who disagree need it forced down their throats.

Jack Waterford

Rodney Tiffen

On the Saturday when the *Australian* devoted two pages to rebutting Robert Manne's *Bad News*, its headline story, labelled exclusive, was "Abbott's plan for northern foodbowl: dam network to be built." There was no comment from anyone outside the Liberal Party – no other politicians, experts, lobby groups or bureaucrats. There was no attempt to give any historical context, no reference to the decades of similar proposals and attempts. There were no costings, only sketchy details of locations and purposes (beyond doubling Australia's agricultural output), nothing to suggest it deserved the splash the paper gave it.

The article resembled the "exclusives" which often seem the stock-in-trade of the press gallery – although usually denied by the journalists involved – where a story is given to one news organisation on the tacit understanding that critical reaction will not be sought and that it will be given prominence.

This is another *Australian* exclusive which is likely to remain exclusive. It is a reminder that, for all its bluster, it is a newspaper often on the brink of mediocrity. When Ken Inglis reviewed the *Australian* after its first year in existence – as he lavished praise on the new paper's contribution to the nation's journalism – one of his criticisms was that the paper was too prone to claiming exclusives.

The founding and growth of the *Australian* in 1964 led directly to an improvement in the rest of the Australian press. While Rupert Murdoch is often, rightly, given credit for his boldness and commitment in founding the national daily, he also very nearly destroyed it. When he forced it to lurch to the right in 1975, and stay there until the advent of the Hawke government in 1983, he severely damaged the paper's credibility with consequences which went straight to the bottom line. Circulation dropped from 136,000 in 1972 to 119,000 in 1982, in a decade when the rest of the quality press substantially increased circulation. Only Murdoch's deep pockets and determination protected the paper from the commercial disaster his politics had forced on it.

Under Chris Mitchell's editorship, however, the paper has managed to combine a strong right-wing ideology with relatively good circulation. Its print circulation is now around 157,000 (combining weekly and weekend sales) and has remained broadly constant over the last decade in contrast to some decline in most Australian newspapers.

The paper's appeals are that it has more extensive international coverage than any other Australian newspaper, a substantial business section, and good reviews and magazine features on a Saturday, as well as voluminous Canberra coverage. Especially in the capital cities outside Sydney and Melbourne, it probably appeals to those whose only local paper is a parochial tabloid, and probably sells well to those who have moved interstate. My guess is that its increasingly pronounced right-wing ideology has little impact on circulation – attracting and repelling readers in broadly equal numbers, and being not terribly relevant to the main strengths of the paper for most of its readers.

It has not had the advertising support that metropolitan dailies derive from retail and many types of classified advertising, but being the only general national paper gives it some compensating advertising advantages. Ironically, the most vociferously "small government" paper in the country is the most dependent on government advertising.

The *Australian*'s distinctive market position partially, but not fully, explains its often idiosyncratic news priorities. As a national paper, it is loath to lead with a story principally of interest to only one state or city, and this often leads it to give prominence to Canberra stories to which other papers accord less importance. Because it is often a second paper for readers, it seeks to avoid duplicating the news agenda of its competitors. Sometimes this leads to innovation and diversity, but sometimes it means that the paper simply fails to follow up or give due prominence to stories where it is off the pace. So the *Australian* was slow to cover the Securency scandal, probably just because the Fairfax papers were so far ahead. But the discretion afforded by its unique position and priorities also easily becomes a licence to push its own agenda.

Manne's essay richly documents the peculiarities and partiality of its coverage of several issues. In the two areas which I have followed most closely – the paper's coverage of the Iraq War and of global warming – Manne's criticisms are fully justified, and in both cases many more examples could be adduced.

The parade of false claims in 2002–03 regarding Saddam's possession of weapons of mass destruction – Iraq importing aluminium tubes for nuclear purposes, importing uranium from Niger, the claims by the defector "Curveball" of biological weapons, meetings between Iraq and al-Qaeda – and the alleged race against time for the West to forestall his imminent aggression constitute one of the most remarkable propaganda exercises in modern democracy. Responsible newspapers such as the *Washington Post* and *New York Times* later reflected publicly on their journalistic failings during this period.

Against initial expectations, the easy defeat of Saddam did not lead to peace in Iraq, but to a disastrous, draining and bloody series of conflicts. Honest, intelligent publications such as the *Economist*, which had supported the invasion, later shared with readers that the war had taken a course that went against expectations, though it continued to give qualified support while openly acknowledging the political and humanitarian costs of the path it had advocated.

The *Australian* did none of this. Its dominant tone editorially and in news selection before the war was intolerant of anyone who questioned the evidence of WMD or the wisdom of invasion. When it looked as if the defeat of Saddam would signal a quick triumph, the paper was rhapsodic. For Greg Sheridan, "The eagle is soaring. The bald eagle of American power is aloft, high above the humble earth, and everything it sees is splendid. For as it soars and swoops it sees victory, power, opportunity."

As the war soured, as the WMD failed to materialise, the paper became increasingly querulous. Its view seemed to be that it had been wrong for the right reasons and the war's critics right for the wrong reasons. As Manne makes clear, the invasion of Iraq and its aftermath was not the *Australian's* finest hour.

The failings of the *Australian* were shared by the Murdoch press around the world and stemmed directly from the crude views of Murdoch himself. His most famous quote in the countdown to war was the prophecy that the overthrow of Saddam would lead to oil costing only US$20 a barrel. Manne quotes an equally absurd claim, with Murdoch saying in February 2003, "We can't back down now, where you hand the whole of the Middle East over to Saddam." It is alarming that a man controlling an empire charged with informing the public really thought that the choice in 2003 was between military invasion and Saddam taking over the whole Middle East.

In some ways the other issue is more intriguing. Murdoch has declared the reality and importance of anthropogenic global warming, but the *Australian* – as well as his news outlets in Britain and America – has continued to give succour to deniers and has sought to highlight all weaknesses in the political or scientific cases for action to reduce it. Two cases where I followed the paper's coverage in some detail were the "Glaciergate" and "Climategate" episodes.

"Glaciergate" concerned the inclusion of a palpably false claim about the rate at which the Himalayan glaciers were melting in the Intergovernmental Panel on Climate Change's report on global warming. The *Australian's* first article was a good, hard-hitting report by Peter Wilson, based principally on an interview with Canadian scientist Graham Cogley, the man who exposed the error and presciently commented that it would be used by global warming sceptics to throw doubt on the whole document. Wilson dissected how the error occurred and the initially inadequate responses by officials to its exposure.

His own newspaper, however, was subsequently guilty of exactly what the article warned against. In the six weeks after that story the *Australian* ran forty-two stories that mentioned glaciers. All referred to the IPCC's Himalayan error; in thirty-eight that was the only reference to glaciers. Only one subsequent article, a column by Mike Steketee, gave any sort of overview of the actual state of the world's glaciers. A report released in that period, which showed glaciers melting at an accelerating rate, was only briefly covered in the *Australian* late in a story dominated by the Himalayan error and headlined "Climate science on thin ice."

The glacier story came some months after one of the most consequential leaks in recent history, when in November 2009, in the lead-up to the Copenhagen summit, thousands of emails were hacked from the Climate Science Unit at the University of East Anglia.

The outraged rhetoric flowed, as climate change deniers pounced on what they saw as evidence of conspiracy. The climate unit's head was forced to stand down and received death threats. Three inquiries were set up in the United Kingdom, with another in America. All of them found that the integrity of the science was intact and that the emails did not place in question any of the published evidence. Manne quotes an *Australian* editorial charging that the emails showed "gaps in the data" and depressing "evidence of sloppy science." In fact both of these charges were refuted by the inquiries, but the newspaper never revisited its judgment.

It is a common charge that the media prominently publishes allegations but gives less coverage to the prosaic facts that later disprove them. In this case the disproportion was particularly stark – by my count at least twenty times as much coverage was devoted to the allegations as to the findings of the inquiries. What-ever its editorial pronouncements on global warming – and Manne does a good job of dissecting their inconsistencies – its news priorities have been irresponsi-bly askew. Its sense of proportionality and representativeness, of what and how to report, is very influenced by its own agendas.

At least most of the time, what's reported in the news columns – with some egregious exceptions, such as its coverage of the report on the Building the Education Revolution program – is broadly accurate. This is more than can be said for much of the writing of its commentariat.

Of course, several of its regular columnists are still essentially reporters whose columns are vehicles for their well-researched analysis of issues and events. But there are others whose appeal is primarily ideological. For these accuracy comes second.

Some years ago I read a Janet Albrechtsen column which particularly annoyed me as ideologically convenient simplification. The column was somewhat eccentric, although not uninteresting, in arguing that Tony Abbott should take as a model Joe Lyons, who resigned from the Scullin Labor government before the 1931 election. It then informed us that Lyons "opposed the [Labor] government's inflationary, pro-Keynesian policies."

The term Keynesian was not then in general use. Keynes was already a very well-known essayist and prominent economist in 1931 (who supported the orthodoxy that Australia should honour its debts – principally run up by preceding conservative governments – and not default on them). But his *General Theory* was published only some years later. Most importantly, in 1931 Australia's GDP had just declined by 10 per cent, and unemployment was approaching its 1932 peak of 29 per cent. It took a man of real courage to oppose "inflationary" policies in that situation, especially because at the time conditions more resembled (what would now be called) a deflationary cycle.

Is there any market punishment for such a distorting simplification that, however, affirms the prejudices of Albrechtsen's constituency? The *Australian* needs to replenish the reporting roots of its journalistic culture. It should seek to raise the research standards of its commentariat. It should seek to report accurately views with which it disagrees. It should make its news selection less "agenda-driven." It should build up and further develop the skills of its staff, its most important asset.

Rodney Tiffen

Ian Lowe

Robert Manne's analysis gives solid quantitative evidence for something critical readers have long suspected. The coverage of climate change by the *Australian* has been outrageous. It breaches the Press Council guidelines and probably its own code of conduct. It regularly blurs the line between news and opinion, consistently misrepresents the science and defends the publishing of uninformed nonsense by claiming this gives "balance." The nadir was a front-page illustrated story about a Sydney surfer who hadn't noticed any sea-level rise, suggesting that his anecdote trumped decades of painstaking observation of tide gauges and detailed analysis of the data by scientists.

At one level, this is clearly an organised campaign to spread misinformation. A scientific colleague of mine was shocked to overhear a discussion between some senior editorial staff of the paper and a recently retired Liberal politician, openly colluding to refine the strategy of spreading doubt about the science. He said that what he witnessed seemed straight from the pages of Naomi Oreskes and Eric Conway's book *Merchants of Doubt*, which documents the systematic promotion of misleading ideas to protect vested interests like the coal industry.

At another level, it is entirely consistent with the general approach of the *Australian*. When CSIRO conducted a major research project on impacts of population growth in 2002, a chorus of the paper's columnists attacked the study. The project's leader, Dr Barney Foran, described as "particularly nifty" the way one journalist, Paul Kelly, took elements of Foran's views and juxtaposed them with other comments out of context to misrepresent the findings. That column and two others were a blatant hatchet job, intended to leave readers with the impression that the research conclusions were not solidly grounded in economic and social reality.

On a recent trip I was provided by Qantas with a copy of the *Australian*. It quoted Robert Manne as saying that he did not believe it made sense to print the opinions of non-scientists on scientific issues, then twisted this to "Newspapers should not publish the opinions of average Australians, reporting only the views of a 'core' of experts on contentious debates, says Robert Manne." The same page-three article modestly stated that the paper "prides itself on providing readers with multiple perspectives on key public issues." On page seven, it reported that the Coalition was being challenged to pull back further from its response to climate change by a rump of MPs who were planning to meet with "Institute of Public Affairs director John Roskam and IPA climate change expert Tim Wilson." The IPA's website shows that it doesn't just deny the science of climate change, but also the link between smoking and lung cancer, even the need to return water to the Murray River! The *Australian* obviously thinks it news-worthy that a group of MPs is being briefed by ideologues with no expertise.

The house ideology is clear: the market is sovereign, the private sector can always be trusted, growth is an unalloyed good, environmentalists are either misguided or malevolent, governments should only interfere in the economy to provide whatever support the private sector wants. Coalition governments are obviously preferred, but ALP governments are acceptable so long as they pursue what are essentially Coalition policies. Economic reform consists of winding the clock back at least to pre-Keynesian times, ideally to the nineteenth century when essential services were controlled by private entrepreneurs who could hold the public to ransom, employers could treat workers with contempt without any interference from unions, and there was no thought of holding back profitable ventures which lay waste to the environment or people's lives.

In 2004 I encountered Paul Kelly myself as part of a media panel at a Canberra symposium on population, economy and environment. He expounded the Murdoch house line, that population growth is good for the economy and there-fore should be supported. In the discussion period, Dr John Coulter pointed out that analysis of OECD data showed there is actually a negative correlation between population growth and such wealth indicators as GDP per head. Kelly said that he hadn't seen the OECD data, but the conclusion didn't make sense to him, and he was sure the people he talked to would not agree with it either! In other words, faced with a choice between the data and the house ideology, he stuck with the ideology. With no apparent sense of irony, his recent piece defending the *Australian* against criticism accused Manne of wanting to impose some sort of orthodoxy and suppress dissenting views.

There can be no doubt about Manne's main point. The *Australian* is a news-paper with an ideological agenda. To advance that agenda, it fabricates "news," publishes fringe views that are intellectually indefensible and largely ignores evidence or opinion that is contrary to the house line. It blurs the line between reportage and editorialising, and it fails to inform its readers of solid science. In a session at the Melbourne Writers' Festival last year, a senior journalist offered the view that this is a commercial decision: being unable to compete in Melbourne with the *Age* or in Sydney with the *Herald* for solid reporting and informed comment, he suggested, the paper has chosen to aim at a different market, positioning itself as "the Fox News of local publishing." I remember what a breath of fresh air the *Australian* was when it first challenged the stuffy Fairfax press of the 1960s. The decline to its current state is very depressing. We deserve better from our only national newspaper.

Ian Lowe

## Robert Manne

I thank Jay Rosen, Tim Flannery, Mark Latham, Jack Waterford, Rodney Tiffen and Ian Lowe for their varied and interesting contributions to this discussion. I have already written more than 40,000 words on the *Australian*. Accordingly I will restrict myself to commenting as briefly as possible on Nick Cater's contribution and even here not on his ad hominem attacks but on his defence of the *Australian*.

In the course of preparing the *Quarterly Essay* I conducted about twenty interviews – some on the record, some not. In every interview I learned things. In almost no case were those interviewed quoted directly. Only the *Australian's* editor-in-chief, Chris Mitchell, was aggrieved. The story of the circumstances surrounding this interview is revealing. I asked for an interview with Mitchell. He offered instead interviews with himself in the company of the editor-at-large, Paul Kelly; with the key members of the paper's Canberra press gallery, Dennis Shanahan and Matthew Franklin; with Nick Cater, who had strategically placed a copy of a book by Richard Boyer, one of the liberal fathers of the ABC, on his desk; with Michael Stutchbury, the economics editor; and with Clive Mathieson, the present editor. In addition, Mitchell presented me with a dossier (mainly of print-outs with which I was already familiar) on subjects I had let him know I intended to discuss. I am not sure what Mitchell thought all this excess would achieve. By the time I visited Holt Street I had read thousands upon thousands of relevant articles. More importantly, I had read Mitchell's *Australian* almost every day for nine years. Clearly he (and Cater) thought a day at Holt Street might suddenly overturn views based on six months' solid research and nine years' reading and reflection. Of course it could not. I requested an interview with Mitchell in part as a matter of courtesy and in part in the hope of picking up something about the atmosphere at the paper's headquarters. I brought a witness, Morry Schwartz. This was prudent. When *Bad News* appeared, an editorial argued that we had offered the paper extract rights, a false claim that later required a "clarification."

As was made clear in *Bad News*, I grounded the critique offered of the *Australian* in seven concrete subject areas so that the essay might transcend generalities. Cater complains, ludicrously, that I did not deal with the paper's entire contents. As he writes: "The essay's disclaimer – 'Many aspects of the paper are not analysed' – admits its restricted scope." He argues that a "more consequential study" would have analysed everything published over forty-seven years. To achieve this, I would have been required to read over 500,000 broadsheet pages. Alternatively, Cater complains that I deal "only" with Aboriginal history, Iraq, climate change, and neoliberalism versus social democracy and that I deal with them because it is on these issues that my personal reputation is staked. This is a perverse way of seeing things. I have become intensely interested in these issues, and in *Bad News* have analysed the *Australian's* performance in regard to them, because they have been, by any objective standards, among the most important political-ideological questions of the past decade in Australia. My reputation or indeed anyone else's is not the issue.

What follows is a brief examination of Cater's critique of the seven subject areas analysed in *Bad News*.

1. *The Making of Keith Windschuttle.* I criticised the *Australian* for its decision to make a book as bad as *The Fabrication of Aboriginal History* the subject of protracted national debate. I did so for two main reasons. Firstly, I regard Windschuttle's book as ultra-reactionary. It describes a people, largely destroyed by British settlement in the space of thirty years, as common criminals and as "agents in their own demise." What does Cater say on this question? Not a word. Secondly, I regard Windschuttle's book as bad history. The claim that it is certain that a mere 120 indigenous people were killed in the course of settlement rests on two absurd ideas: namely that no one died of wounds and that for every indigenous death there must exist an extant documentary record. What does Cater say on this issue? Not a word. He defends the *Australian's* record here solely on the grounds of "pluralism." This defence is empty. If the Fairfax press decided to make a book alleging that Israeli agents were responsible for 9/11 the subject of a prolonged national debate, that could also be defended on Cater's grounds of pluralism.

2. *The Invasion of Iraq.* Like all 175 Murdoch newspapers, the *Australian* vigorously prosecuted the case for the invasion of Iraq. It did so in a manner that treated the opponents of the war, especially those it considered "left-wing," with contempt. This contempt was strongest in the paper's editorials and in the commentaries of its foreign editor, Greg Sheridan. The *Australian* justified the invasion on the basis

that there could be no doubt about Iraq's possession of WMD and that there could be no question that Saddam Hussein would use these weapons either to dominate the Middle East or to provide such weapons to Islamist terrorists. It regarded the opposition of the UN Security Council to the invasion as irrelevant. Following the invasion, no WMD were found. Since the invasion, up to 400,000 people have lost their lives. Millions more have been displaced or become refugees. Yet unlike several newspapers, the *Australian* has uttered not one word of apology or even explanation to its readers for its arrogance and misjudgment. How then does Cater respond to this case? He claims that all this is simply a matter of opinion. For him, my argument about the paper's appalling coverage of Iraq and its aftermath boils down to this: "He does not agree." Beyond this, his only quibble is that I analyse the Iraq performance of Sheridan rather than Paul Kelly. Cater quotes Kelly on Howard in March 2003: "He has failed to mount a persuasive argument that a war to disarm Iraq is an imperative now ..." In December 2002 Kelly, however, argued: "It is ... fatuous to think Australia could refuse to make any military contribution [to the invasion] without hurting the alliance ..." My essay was an attempt to avoid cherry-picking of the kind Cater advocates. I remember Kelly's inconsistent performance over Iraq only too well. He was all over the shop. On the Iraq War, Sheridan was anyhow a more important voice than Kelly.

3. *Media Watch. Bad News* analysed the *Australian's* many brain snaps with regard to criticism mounted by this ABC program. The point I was trying to make was that while the *Australian* is only too willing to dish out criticism on a daily basis, when in turn the paper is criticised it explodes with rage and indignation. Cater avoids the issue. As it happens, the manic way the paper responded to *Bad News* is a good example of what I had in mind. A grown-up paper would have taken the *Quarterly Essay* in its stride. Instead, in a month, the *Australian* published 25,000 words of attacks: columns, reports, editorials, letters to the editor, "Cut and Paste" items and a cartoon showing me on the toilet, shitting. At first I was refused a right of reply. Only after I wrote to John Hartigan, the head of News Limited, informing him that I intended to take the case to the Press Council did it offer me 1000 words. Cater thinks the culture of the *Australian* is "liberal." He seems not to know what the word means.

4. *Climate Change.* In *Bad News* I argued that the *Australian's* performance over the issue of climate change represented a "truly frightful hotchpotch of ideological prejudice and intellectual confusion." Cater's attempted defence of his paper's performance on climate change is an excellent example. Chris Mitchell in December 2010 defended his paper's performance over climate change with two

arguments. Firstly, he claimed that the paper's editorials consistently supported the conclusions of the climate scientists. I analysed more than 300 editorials showing this to be a falsehood. What does Cater say about this? Not one word. Mitchell then argued that his paper, on balance, had consistently argued in favour of action on climate change. To decide whether this was so, I analysed some 880 feature articles and opinion columns on climate change published between January 2004 and April 2011. I outlined in the essay what in my view counted as coverage favourable and unfavourable to action on climate change. This analysis showed that articles published in the *Australian* were opposed to action on climate change by a ratio of four to one. Cater simply cannot follow the argument. He claims I did not allow for what he calls "peer review" of my work. In fact, in the notes to the essay I published the Factiva database formula used to generate the articles. He claims that I discovered only twenty-nine "denialist" articles in the *Australian*. This is quite false. The twenty-nine articles refer only to the "denialist" articles written by those boasting scientific credentials. In *Bad News* I did not offer any figure for "denialist" articles written by non-scientists. If I had added together the denialist articles written by scientists and non-scientists – Christopher Pearson alone contributed more than twenty examples of the latter – the figure would probably have exceeded 100. Finally, Cater claims that if I had taken up his February 2009 offer to write for the *Australian* I could have more than corrected the *Australian's* imbalance by writing on climate change once every six weeks. The imbalance between articles favourable and unfavourable to action on climate change was 180 to 700. To correct the imbalance I would have had to write 520 articles in twenty-six months or almost one every day.

5. *The Rise and Fall of Rudd. Bad News* analysed the *Australian's* performance between Rudd's selection as leader of the Opposition and his loss of the prime ministership in June 2010 following a caucus coup. Two main criticisms were raised. I argued that in the way the paper reported the rise and fall of Rudd, Chris Mitchell's volatile personal relationship with Rudd represented a systematically distorting factor. And I argued that following Rudd's February 2009 *Monthly* essay against neoliberalism, which offended the *Australian's* house theology, criticism of his government not only escalated but also became increasingly unbalanced, reaching a mad crescendo in the six weeks of unremittingly hostile front-page coverage of the supposed disaster of the super-profits mining tax. In its truly strange editorials, the *Australian* used the issue of the mining tax to argue that Rudd was not merely as bad a prime minister as Whitlam but unfit to be prime minister at all. Cater's analysis here is deliberately misleading. He pretends I

argue that Rudd was destroyed by two editorials. In *Bad News* I analyse not two editorials but eighteen months of accelerating anti-Rudd hostility. He suggests that I claim that the *Australian* was solely responsible for destroying the Rudd government. In the critical passage of *Bad News* on this point I conclude that there are "two ways" of misunderstanding the role played by the *Australian* in the demise of Rudd. One is "to exaggerate." Rudd was far "too politically isolated." The other is to minimise its impact. Rudd did not lose because of public opinion. The remorseless campaigning of the *Australian* helped to "crystallise" anti-Rudd opinion in the caucus and more generally among the political class. Cater deliberately misleads by leaving out half of the conclusion. Incidentally, in his criticism of Rudd's *Monthly* essay on the failure of "extreme capitalism," Cater informs us helpfully that all American and British governments over the past thirty years have been influenced by what some call "neoliberalism" and others market fundamentalism. Bunkered down in Holt Street, Cater appears not to have noticed either the devastation wrought upon the world by the market fundamentalism and the greed of Wall Street since the collapse of Lehman Brothers in 2008, or the fact that in the United States one of the wonders of neoliberalism is to have produced a grotesque society where the middle and working classes have stagnated for thirty years and 40 per cent of national wealth has trickled upwards into the hands of the top 1 per cent.

6. *Posetti and Behrendt.* In recent days the *Australian* has become the great defender of freedom of speech. Yet Cater thinks it fine that almost a year ago his editor-in-chief threatened a defamation writ against a lecturer in journalism on the basis of an accurate tweet of a former employee's conference talk. The only thing that seems to interest him is that I found the threat "astonishing." He also seems genuinely to think that the best way of reporting on the divisions between indigenous intellectuals over the federal government's Northern Territory Intervention was to seize upon a foolish bad-joke tweet in order to prosecute a three-week-long campaign of vicious character assassination against Larissa Behrendt.

7. *The Greens.* Cater doesn't even try to defend the campaign his newspaper has waged against the Greens since the 2010 election, which is documented in detail in *Bad News* but which his editor-in-chief defended in the pages of his paper as "balanced." Instead he accuses me of writing a conspiratorial version of history when I claim that the present power of the Murdoch press in Australia represents a threat to democracy. My argument goes like this. The Murdoch press has 70 per cent of the circulation of the statewide and national press in Australia. It is owned by an ideologue with a special interest in Australia. Over the past twelve months his newspapers have been campaigning against the Gillard

minority government. The chief executive of News Limited, John Hartigan, has openly stated in an ABC television interview that his "company" has a problem with the "leadership vacuum" represented by the Gillard minority government about which something needs to be "done." Earlier this year, the key News Limited editors and journalists met with Murdoch in California. I was informed by one of the attendees – Chris Mitchell – that on the second day of the meeting Australian politics was discussed. A senior minister I interviewed for *Bad News* was informed by someone in attendance about the anti-Gillard government discussions that took place. Cater is employed by a ruthless corporation that is now known to have hacked into hundreds of mobile phones in the United Kingdom. It will take much more than his mockery to convince Australian democrats that we have no grounds for concern.

Cater ends his piece by noting the "bad grace" with which I have turned down his paper's February 2009 invitation to contribute. He quotes a comment I made at Byron Bay about not wishing to give the *Australian* legitimacy by being published in it. What I meant was that I had no intention of being wheeled out every time the paper was attacked for being so right-wing. At the time I suspected but was not sure that providing an alibi was the real reason for the unexpected invitation. During the course of the controversy over *Bad News*, the editor-in-chief of the *Australian*, Chris Mitchell, advised Fairfax not to publish my "crap." What I had suspected about the invitation was in this way unintentionally confirmed. Nick Cater's earlier courtesy and even flattery, which I accepted at face value, had been nothing but a hypocritical mask.

Robert Manne

**Nick Cater** has been editor of the *Weekend Australian* since 2007 and a senior editor at the *Australian* since 2004. He edited the 2006 book *The Howard Factor*.

**Andrew Charlton** was senior economic adviser to the prime minister from 2008 to 2010. During that time he served as Australia's senior official to the G20 summits and the prime minister's representative to the Copenhagen Climate Conference. He previously worked for the London School of Economics, the United Nations and the Boston Consulting Group and received his doctorate in economics from Oxford University, where he studied as a Rhodes Scholar. He is the author of *Ozonomics* (2007) and *Fair Trade for All* (2005), co-written with Nobel Laureate Joseph Stiglitz.

**Richard Flanagan**'s novels are published in twenty-six countries. His most recent book is a collection of non-fiction writings, *And What Do You Do, Mr Gable?*

**Tim Flannery** has published over a dozen books, including *The Future Eaters*, *The Eternal Frontier*, *The Weather Makers*, *Now or Never: A Sustainable Future for Australia?* and *Here on Earth*. He was Australian of the Year in 2007.

**Mark Latham** is a former leader of the Australian Labor Party and was Opposition leader from 2003 to 2005. He writes a regular column for the *Australian Financial Review* and his books include *Civilising Global Capital* and *The Latham Diaries*.

**Ian Lowe** is an emeritus professor of science, technology and society at Griffith University and president of the Australian Conservation Foundation. His books include *A Big Fix*, *Living in the Hothouse* and *A Voice of Reason: Reflections on Australia*.

**Robert Manne** is professor of politics at La Trobe University and a regular writer for the *Monthly*. His recent books include *Making Trouble: Essays Against the New Australian Complacency* and *Goodbye to All That?: On the Failure of Neo-liberalism and the Urgency of Change* (as co-editor).

**Jay Rosen** has been part of the journalism faculty at New York University since 1986, serving as department chair from 1999 to 2005. His blog on journalism, *PressThink*, has been running since 2003.

**Rodney Tiffen** is an emeritus professor of government and international relations at the University of Sydney. His most recent book (with Ross Gittins) is *How Australia Compares*.

**Jack Waterford** is editor-at-large of the *Canberra Times*, for which he has worked since 1972.

# SUBSCRIBE to Quarterly Essay & SAVE nearly 40% off the cover price

**Subscriptions:** Receive a discount and never miss an issue. Mailed direct to your door.

☐ **1 year subscription** (4 issues): $49 a year within Australia incl. GST. Outside Australia $79.

☐ **2 year subscription** (8 issues): $95 a year within Australia incl. GST. Outside Australia $155.

\* All prices include postage and handling.

**Back Issues:** (Prices include postage and handling.)

☐ **QE 1** ($10.95) Robert Manne *In Denial*
☐ **QE 2** ($10.95) John Birmingham *Appeasing Jakarta*
☐ **QE 4** ($10.95) Don Watson *Rabbit Syndrome*
☐ **QE 5** ($12.95) Mungo MacCallum *Girt by Sea*
☐ **QE 6** ($12.95) John Button *Beyond Belief*
☐ **QE 7** ($12.95) John Martinkus *Paradise Betrayed*
☐ **QE 8** ($12.95) Amanda Lohrey *Groundswell*
☐ **QE 10** ($13.95) Gideon Haigh *Bad Company*
☐ **QE 11** ($13.95) Germaine Greer *Whitefella Jump Up*
☐ **QE 12** ($13.95) David Malouf *Made in England*
☐ **QE 13** ($13.95) Robert Manne with David Corlett *Sending Them Home*
☐ **QE 14** ($14.95) Paul McGeough *Mission Impossible*
☐ **QE 15** ($14.95) Margaret Simons *Latham's World*
☐ **QE 16** ($14.95) Raimond Gaita *Breach of Trust*
☐ **QE 17** ($14.95) John Hirst *"Kangaroo Court"*
☐ **QE 18** ($14.95) Gail Bell *The Worried Well*
☐ **QE 19** ($15.95) Judith Brett *Relaxed & Comfortable*
☐ **QE 20** ($15.95) John Birmingham *A Time for War*
☐ **QE 21** ($15.95) Clive Hamilton *What's Left?*
☐ **QE 22** ($15.95) Amanda Lohrey *Voting for Jesus*
☐ **QE 23** ($15.95) Inga Clendinnen *The History Question*
☐ **QE 24** ($15.95) Robyn Davidson *No Fixed Address*
☐ **QE 25** ($15.95) Peter Hartcher *Bipolar Nation*
☐ **QE 26** ($15.95) David Marr *His Master's Voice*
☐ **QE 27** ($15.95) Ian Lowe *Reaction Time*
☐ **QE 28** ($15.95) Judith Brett *Exit Right*
☐ **QE 29** ($16.95) Anne Manne *Love & Money*
☐ **QE 30** ($16.95) Paul Toohey *Last Drinks*
☐ **QE 31** ($16.95) Tim Flannery *Now or Never*
☐ **QE 32** ($16.95) Kate Jennings *American Revolution*
☐ **QE 33** ($17.95) Guy Pearse *Quarry Vision*
☐ **QE 34** ($17.95) Annabel Crabb *Stop at Nothing*
☐ **QE 35** ($17.95) Noel Pearson *Radical Hope*
☐ **QE 36** ($17.95) Mungo MacCallum *Australian Story*
☐ **QE 37** ($20.95) Waleed Aly *What's Right?*
☐ **QE 38** ($20.95) David Marr *Power Trip*
☐ **QE 39** ($20.95) Hugh White *Power Shift*
☐ **QE 40** ($20.95) George Megalogenis *Trivial Pursuit*
☐ **QE 41** ($20.95) David Malouf *The Happy Life*
☐ **QE 42** ($20.95) Judith Brett *Fair Share*
☐ **QE 43** ($20.95) Robert Manne *Bad News*

**Payment Details:** I enclose a cheque/money order made out to Schwartz Media Pty Ltd. Please debit my credit card (Mastercard or Visa accepted).

Card No.

Expiry date       /                                          Amount $

Cardholder's name                          Signature

Name

Address

Email                                              Phone

**Post or fax this form to:** Quarterly Essay, Reply Paid 79448, Collingwood VIC 3066 /
Tel: (03) 9486 0288 / Fax: (03) 9486 0244 / Email: subscribe@blackincbooks.com
Subscribe online at **www.quarterlyessay.com**